The Ultimate PS3™ Repair Guide

Published by:
Longitudes, LLC.
200 E. Inglefield Rd.
Evansville, IN 47725

ISBN 978-0-578-05477-3

Printed in the U.S.A.
First Edition

Contents

Chapter 1 – Introduction

     The information in this book was attained through the experience of the electronics repair technicians at Tri-State Module, Inc. This company is an independent national repair center for all video game systems and high definition TV circuit boards. Tri-State Module, Inc. has been repairing video game systems for over 3 years. With this experience and a 98% repair success rate, they are passing this wealth of knowledge on to anyone who wants to learn. This repair guide is so comprehensive that it covers almost all symptoms and repairs (not just the most common ones). This book was written to show anyone how to repair a PS3™ system regardless of electronic repair background.

     Following the methods in this guide, the chances you will be able to repair your system are extremely high. Most people would be surprised how simple and inexpensive it was to repair their system. The tools used in each repair are common and if parts are needed, they are widely available on the internet for very reasonable prices. Each symptom and repair is explained in detail with full color pictures. In the next chapter, we will start by diagnosing the problem with the system. This guide covers repairs for the older systems only. The newer slim models are not covered here since they are still under manufacturer warranty. Below is a list of the older U.S. models with information about them. Please note the model can be identified by looking on the white sticker located at the back of the system. Both the model and serial number are located on this sticker. Internal part changes were made with each model to make it more efficient, reliable and reduce cost.

Model Number	*Hard Drive Size*	*PS2™ Compatible*	*# of USB Ports*
CECHA01	60 GB	Yes	4
CECHB01*	20 GB	Yes	4
CECHE01**	80 GB	Yes	4
CECHG01	40 GB	No	2
CECHH01	40 GB	No	2
CECHP01	160 GB	No	2
CECHK01	80 GB	No	2
CECHL01	80 GB	No	2

* - The CECHB01 model did not have Wi-Fi built in.
** - The CECHE01 model achieved PS2™ compatibility through software emulation.

Chapter 2 – Problem Diagnosis

To diagnose the problem, you will need to get detailed information on exactly what the system is doing. With this information, you can isolate the part that is causing the issue. Once the defective part has been identified, we can move forward to the disassembly of the system and the repair. So, let us start the diagnosis. On the back of the console above the AC plug-in is a power switch. This power switch is shown in Picture 1 for reference. With the power cord plugged in, turn the switch off. There should be no lights on the front of the console. Wait 30 seconds and turn the switch back to the on position.

Picture 1 (AC power switch on back of console)

At this point, match the symptom that is being experienced with what is described below. This is a comprehensive list of the possible symptoms with their associated chapter on the repair process. Remember to visit Chapter 3 afterward to follow the disassembly process of the system.

Symptom #1 - System will not turn on and does not respond when the power button is pushed. There is no red light on the front of the console.

Diagnosis #1 – The system has an issue that related either to the power supply or the motherboard. Please visit Chapter 7 for further diagnosis and repair procedures.

Symptom #2 – System will turn on, but beeps 3 times and shuts down soon after. A yellow light appears quickly during this shut down process.

Diagnosis #2 - This symptom is often referred to as YLOD (Yellow Light of Death). Please visit Chapter 9 for the repair procedure.

Symptom #3 – System will not read a disc. The disc will not show up in the XMB (menu system) and / or it will start to read followed by an error message about disc operation issues. It may selectively read some discs, but not others.

Diagnosis #3 – The disc drive is defective. Please visit Chapter 4 for further diagnosis and repair procedures.

Symptom #4 – While the system is reading a disc, the video will suddenly freeze. Usually, the system will stop responding also.

Diagnosis #4 – This symptom is caused by a defective disc drive. Please visit Chapter 4 for further diagnosis and repair procedures.

Symptom #5 – A disc is stuck in the drive and it will not eject.

Diagnosis #5 – The disc drive is defective. Please visit Chapter 4 for further diagnosis and repair procedures.

Symptom #6 – The drive will not accept a disc.

Diagnosis #6 – The disc drive is defective. Please visit Chapter 4 for further diagnosis and repair procedures.

Symptom #7 – The XMB (menu screen) never appears on the screen. Only a colored background with a "floating ribbon" shows on the screen.

Diagnosis #7 – This symptom is caused by corrupted files or a defective hard drive. Please visit Chapter 5 for further diagnosis and repair procedures.

Symptom #8 - An error message appears on the screen related to software issues. There are several different messages that pertain to the hard drive such as: "The correct hard disc was not found", "System information or software is corrupted", "File system is corrupted and needs to be restored." and "Database needs to be rebuilt". You may continue the process and it keeps coming back to the same error message in an endless loop.

Diagnosis #8 – This symptom is caused by corrupted files or a defective hard drive. Please visit Chapter 5 for further diagnosis and repair procedures.

Symptom #9 – System will not complete a software update.

Diagnosis #9 – This can be caused by a couple different issues. Visit Chapter 5 for further diagnosis and repair.

Symptom #10 – System will not display video / picture even after confirming all AV (audio/video) cables have been connected properly.

Diagnosis #10 – This symptom can be caused by several different issues. Please visit Chapter 6 for further diagnosis and repair procedures.

Symptom #11 – The system does not output audio. AV connections and audio settings have been verified.

Diagnosis #11 – This is usually caused by an issue on the motherboard. Visit Chapter 6 for repair procedures.

Symptom #12- The system will not connect to the internet or the PlayStation® network.

Diagnosis #12 – This can be caused by multiple issues. Visit Chapter 11 for further diagnosis and repair procedures.

Symptom #13 – A message appears on the screen saying the system is overheating and will shutdown. The system will usually shut down soon after.

Diagnosis #13 – Visit Chapter 8 for further diagnosis and repair procedures.

Symptom #14 – The media card slots do not work correctly. The system will not recognize, read, or write to memory cards when inserted.

Diagnosis #14 – The media card board is defective. Please visit Chapter 7 for repair procedures.

Note: The symptom referred to as "The Red Screen of Death" is not repairable. The characteristic of this symptom is it will display a red screen stating a fatal error has occurred and to contact customer service. If this symptom occurs, please contact Sony for assistance.

Chapter 3 – System Disassembly

Complete Tool Checklist

- Philips head screwdriver
- Small tipped flat-head screwdriver
- Small tipped Philips head screwdriver
- T10 Torx head screwdriver
- Tweezers
- Can of compressed air
- USB flash drive
- Heat gun (not a hair dryer!)
- High quality thermal compound
- Non-corrosive cleaner
- Cotton swabs or cloth
- Paper clip

Listed above is a complete tool checklist for all the repairs in this guide. Most repairs will only use a few of these tools. The corresponding repair chapter for your symptom will prompt you to use what tools are needed. Please remember to thoroughly clean out the system with a can of compressed air as it is disassembled. All of this cleaning will lead to better cooling and a longer life span for the system. Note that depending on the repair, you will be removing ribbon cables from connectors. There is a tab (usually black or brown) to flip up on these connectors that releases the cable. Please be gentle with these tabs as they can break easily. A small tipped flat-head screwdriver works well for this. Clear organization of all parts removed is highly recommended as it will lead to faster and easier reassembly.

Disconnect the power cable from the system before starting disassembly. The first step to disassembly is to remove the warranty seal. This is shown in Picture 2. Please note that any warranty on the console from Sony will be void once the warranty sticker is removed. Peel this sticker off and remove the black plastic insert behind it. This can be pulled off by hand. Behind that rubber insert is a screw that holds the reflective top cover of the system in place. Use a T10 Torx head screwdriver to remove this screw as shown in Picture 3. The cover will slide back and up off of the console. Your system may have a metal tab where the last screw was removed like in Picture 4. Not all systems have this metal tab. Please note if the system has this metal tab and be careful not to lose it as it will be needed for re-installation of the top cover later.

There are 7 screws that secure the top of the system in place as shown in Picture 5. Note that there are arrows pointing to the screws to remove on the system itself also.

The screw in the upper right corner is shorter than the others and is labeled with an "S" next to the arrow on the system. Remove all these screws. Pull the top of the case off from the back as shown in Picture 6. You may have to apply some force to remove it. You may hear tabs popping during this and that is normal.

Picture 2 (Location of warranty seal and hard drive access door)

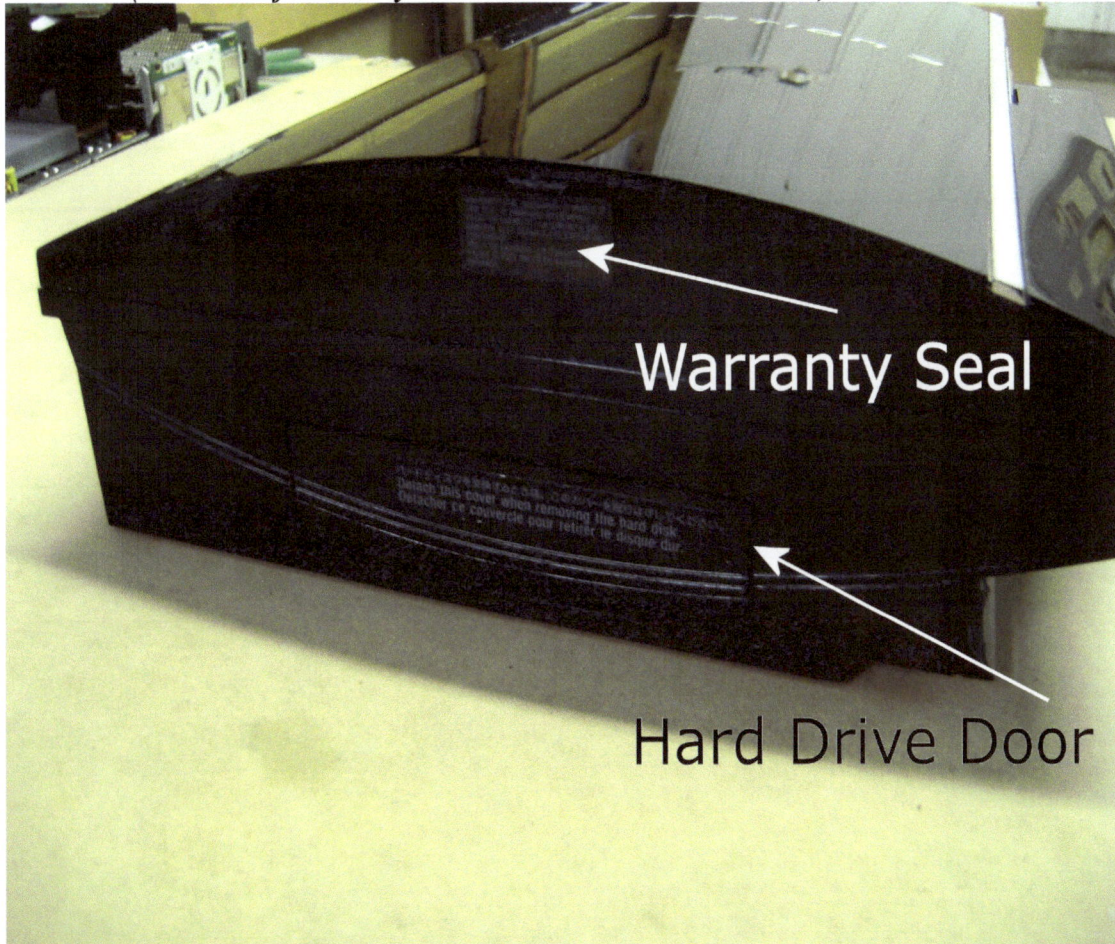

Warranty Seal

Hard Drive Door

Picture 3 (Location of screw for top cover removal)

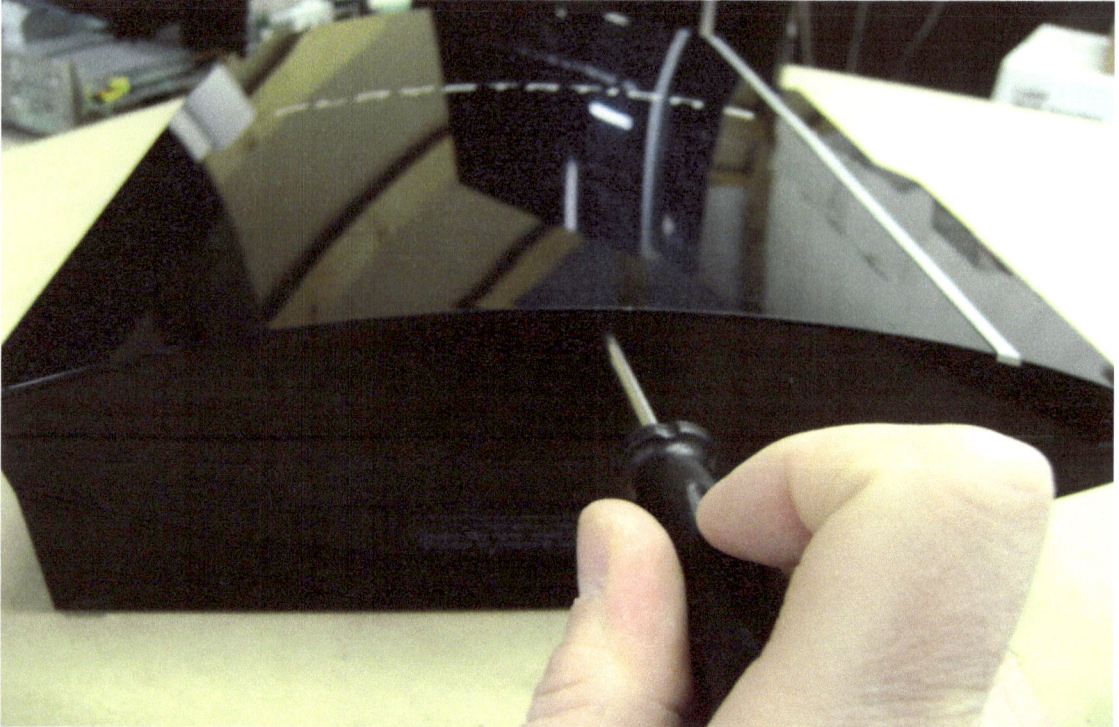

Picture 4 (Metal tab on certain models)

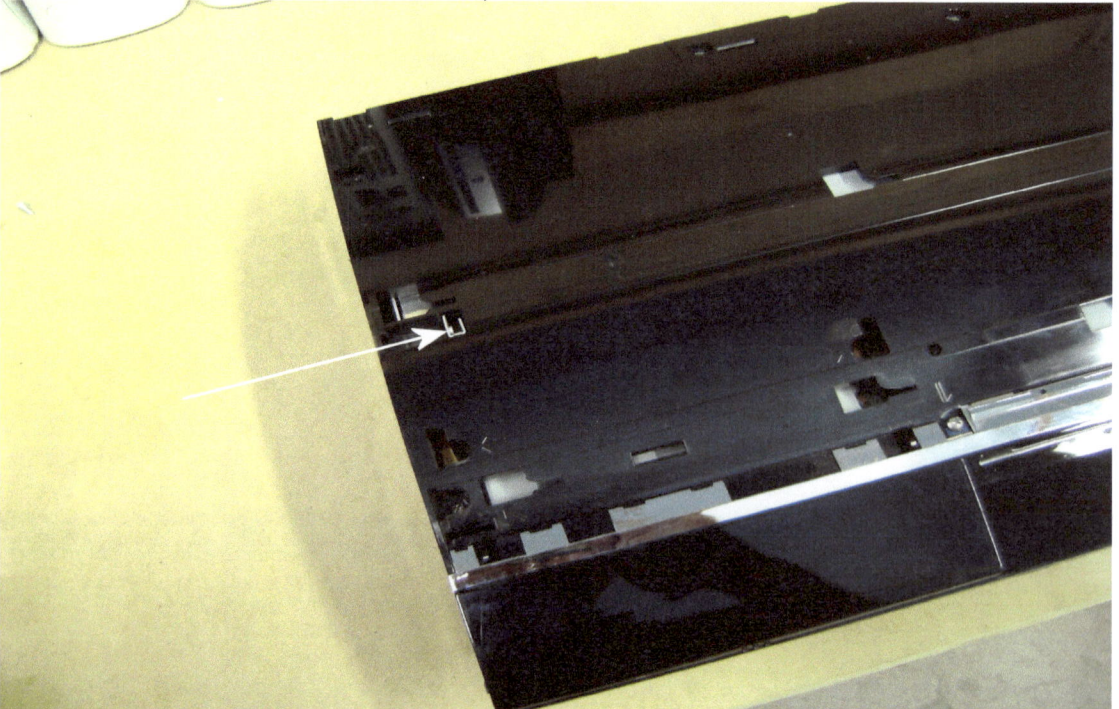

Picture 5 (Location of screws for case removal)

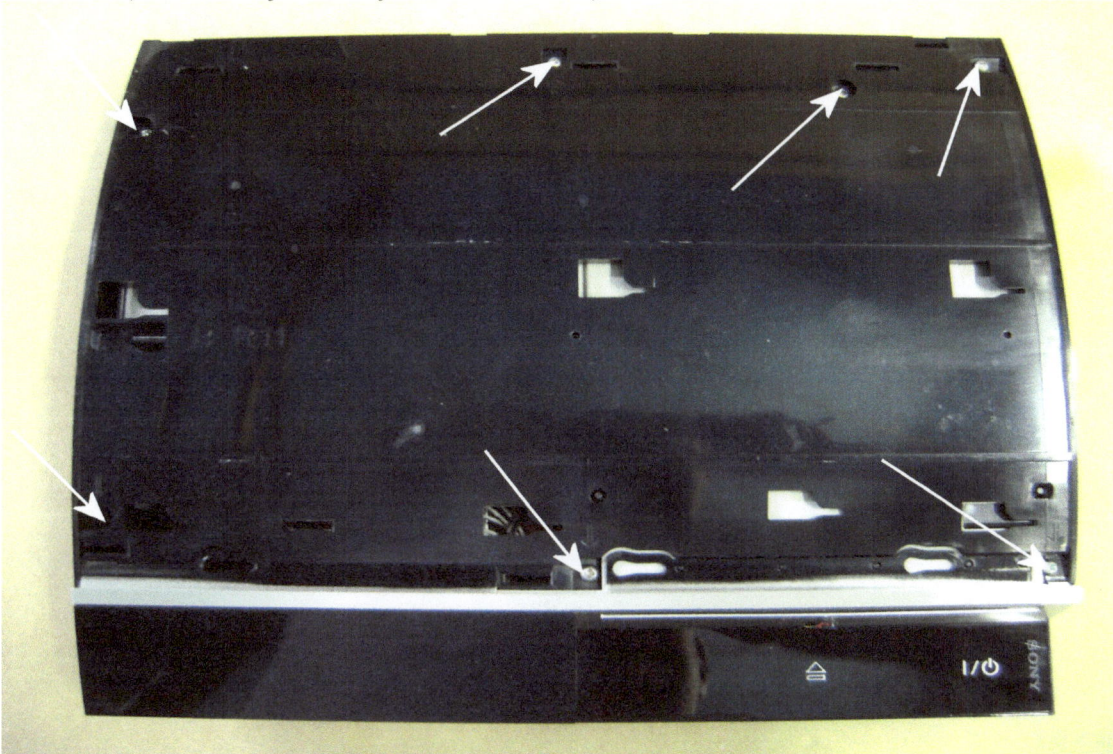

Picture 6 (Removal of top casing)

Now the internal parts of the system are exposed. Several revisions to the internal parts of the system were made with each model change. The 3 main styles are shown in Picture 7, 8 and 9. Your system should look very similar to one of them. Take this time to familiarize yourself with the location and name of the major parts inside. Regardless of model, disassembly is very similar. At this point, you have access to many major parts to start most repair procedures. The next topic in this chapter is complete system disassembly down to the motherboard. Do not proceed any further in this chapter unless instructed to do so according to your diagnosis chapter!

Picture 7 (Identification of major system components version 1)

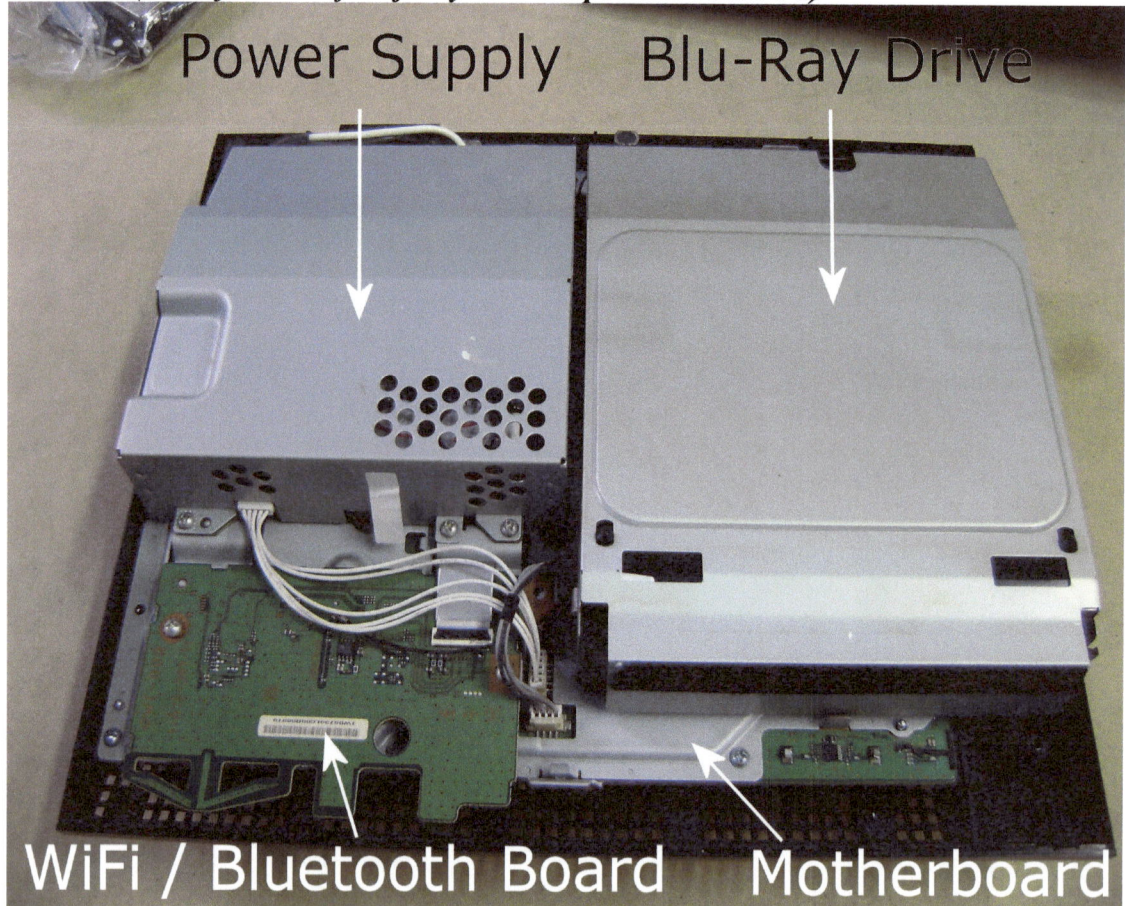

Picture 8 (Identification of major system components version 2)

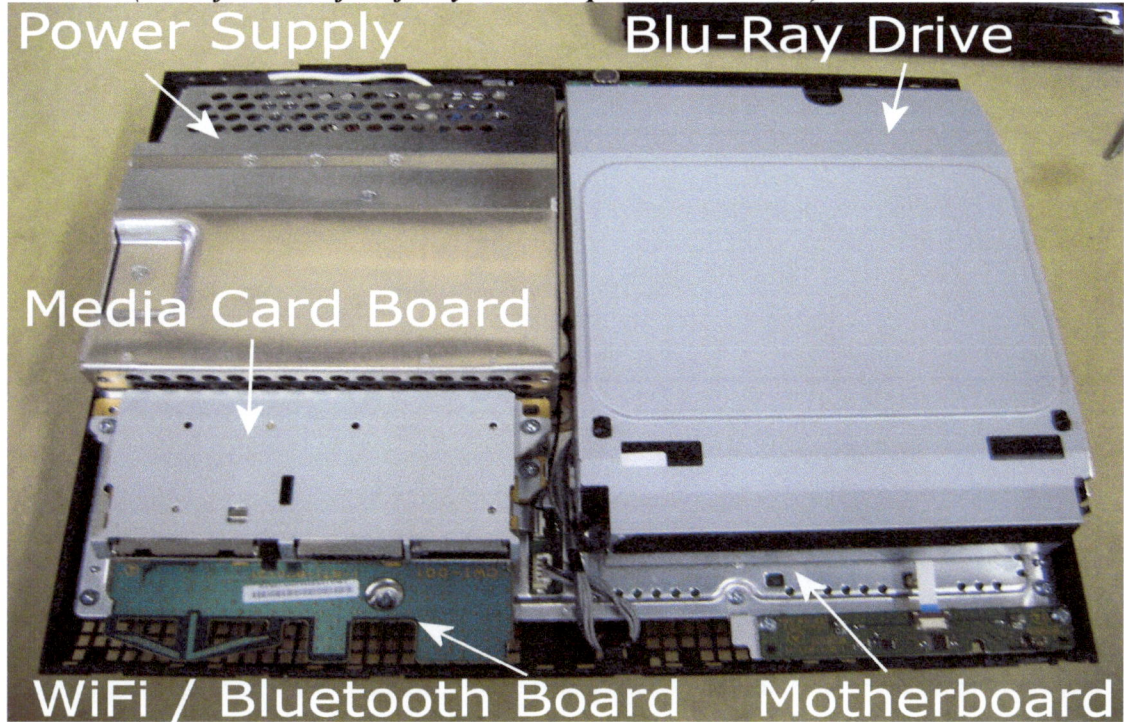

Power Supply

Blu-Ray Drive

Media Card Board

WiFi / Bluetooth Board

Motherboard

Picture 9 (Identification of major system components version 3)

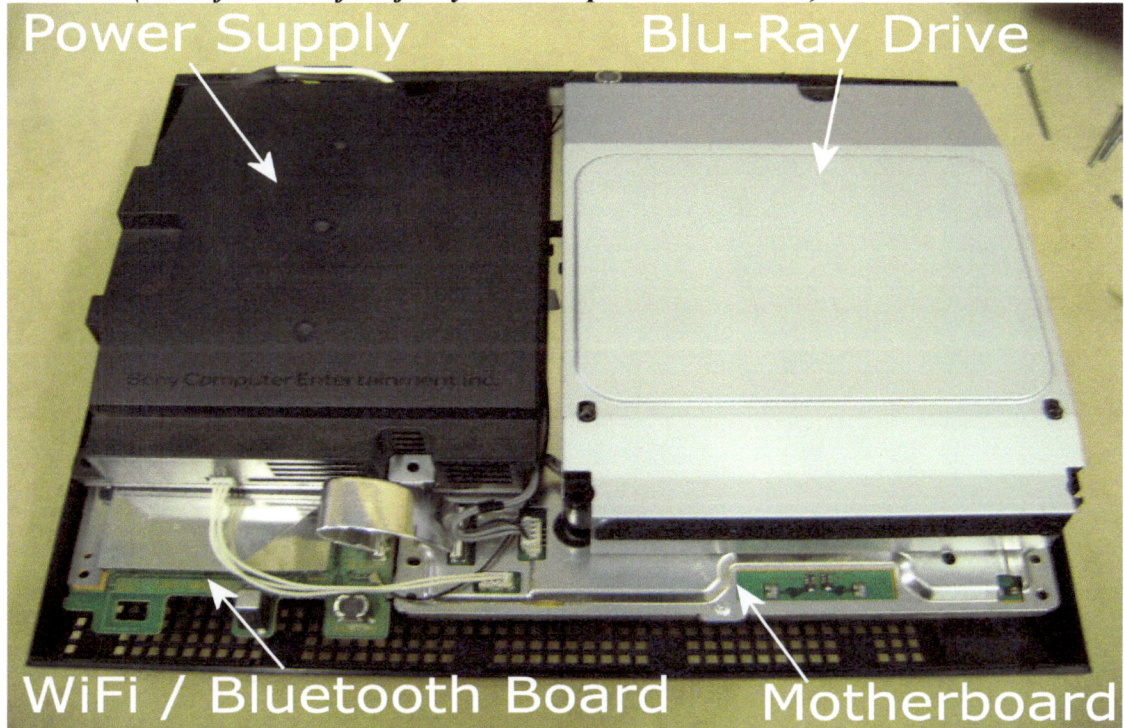

Power Supply

Blu-Ray Drive

WiFi / Bluetooth Board

Motherboard

- Media Card Board Removal

If your system did not look like Picture 8, you can skip this step. The media card board will need to be removed before accessing the power supply. There are two screws for removal and a ribbon cable to disconnect like in Picture 10 and 11.

Picture 10 (Media card board removal if equipped)

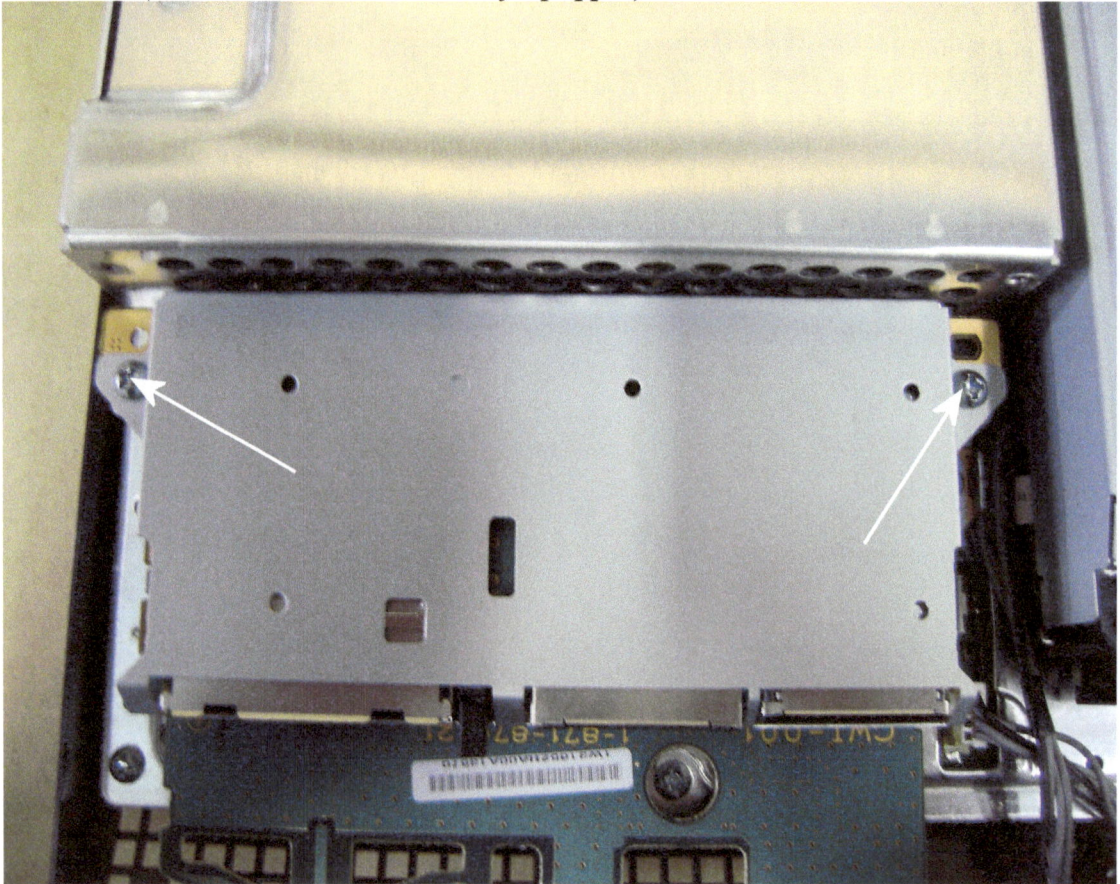

Picture 11 (Media card board removal continued)

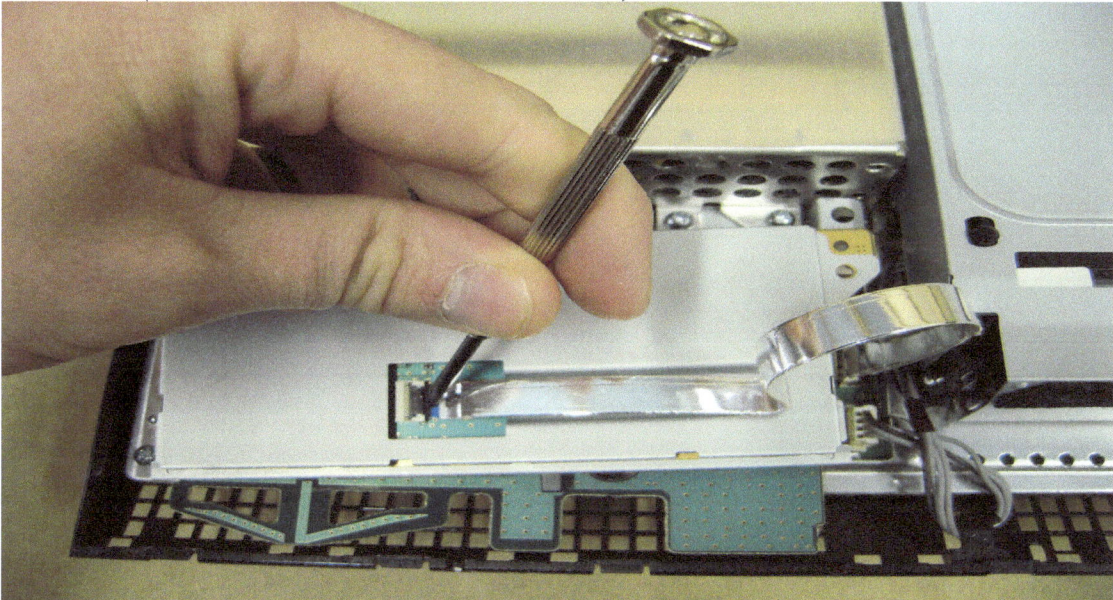

- Power Supply Removal

Disconnect the 2 cables and remove all screws that hold the power supply in place as shown in Pictures 12 and 13. Pull the power supply up off of the board. Your power supply may look different than the one in the picture, but the concept of removal is same for all.

Picture 12 (Identification of parts for removal of power supply)

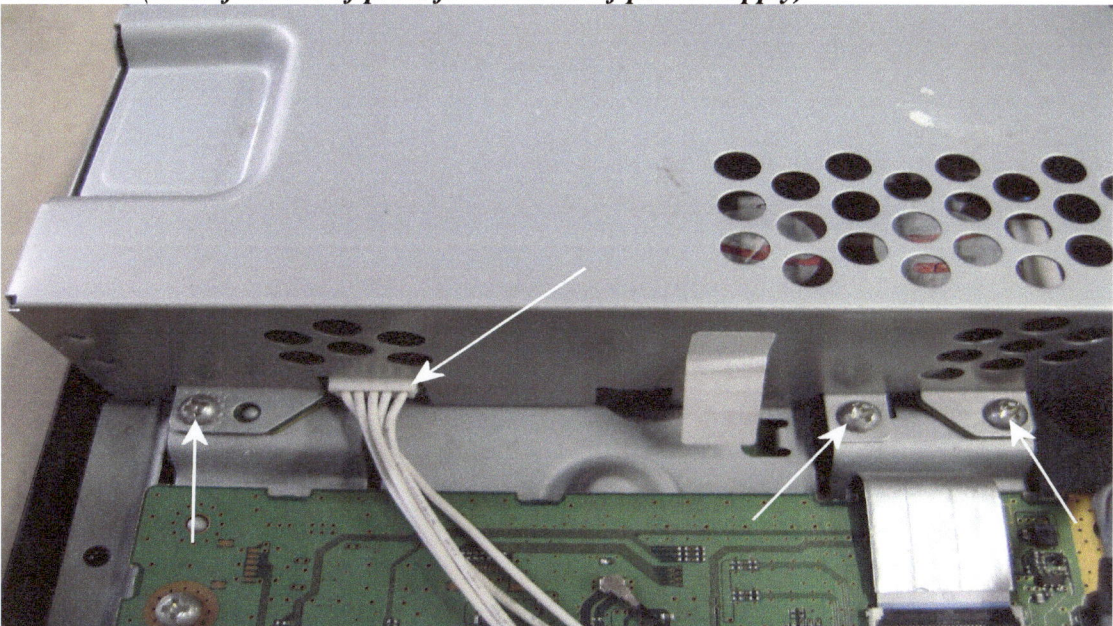

Picture 13 (Identification of parts for removal of power supply continued)

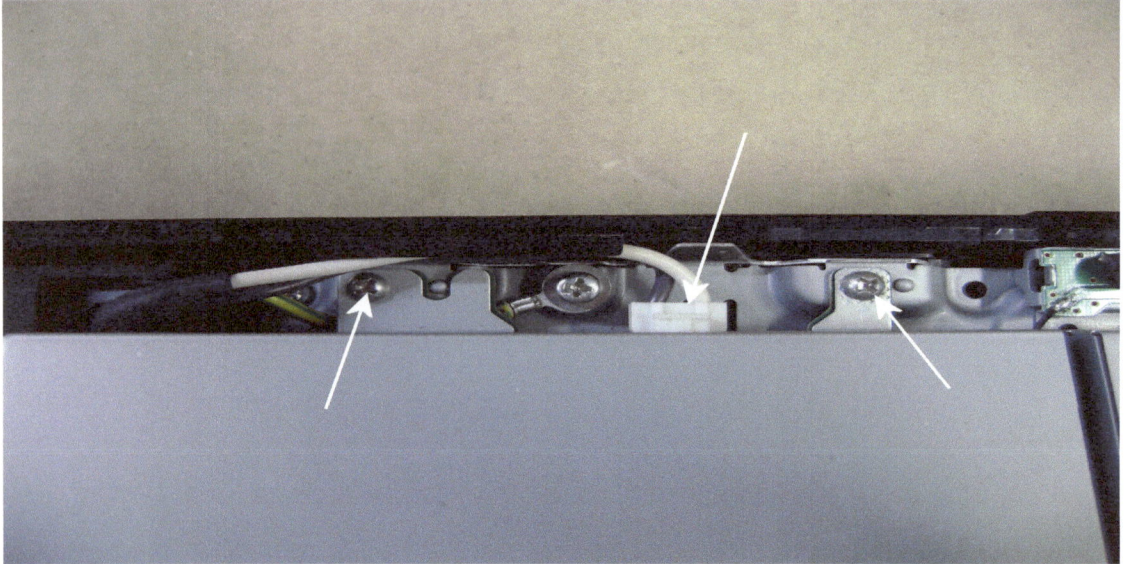

- Disc Drive Removal

Now remove the disc drive by disconnecting the 2 cables (one is on the left and the other is underneath the drive) as shown in Pictures 14 and 15. There may be white tape covering the ribbon cable connection on the bottom of the drive that will need to be removed before disconnecting the cable. Your drive may look different than the one in the picture, but the concept of removal is same for all.

Picture 14 (Cable 1 for disc drive removal)

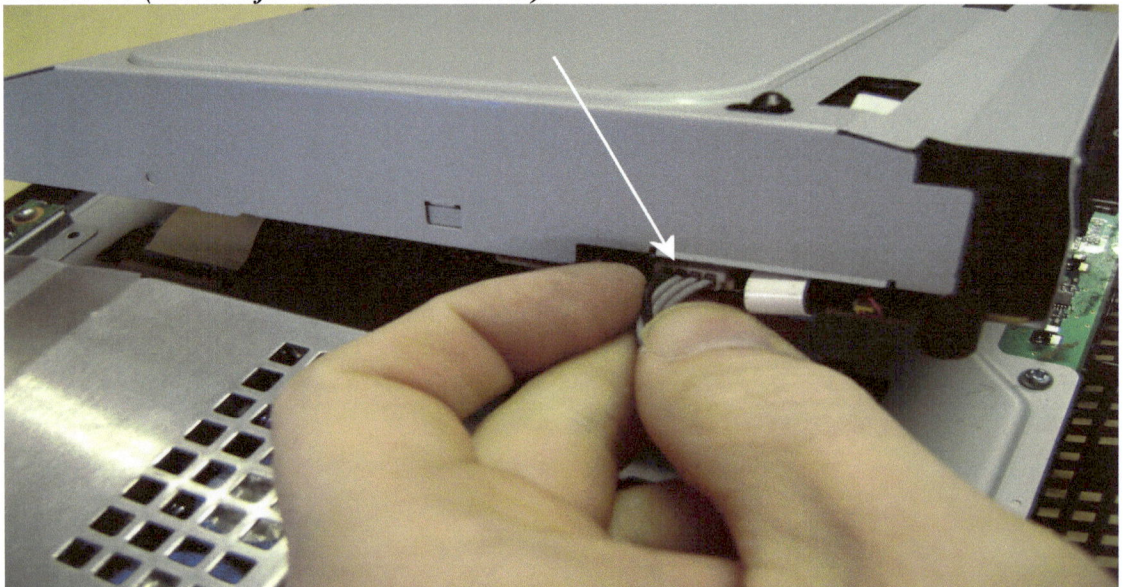

Picture 15 (Cable 2 for disc drive removal)

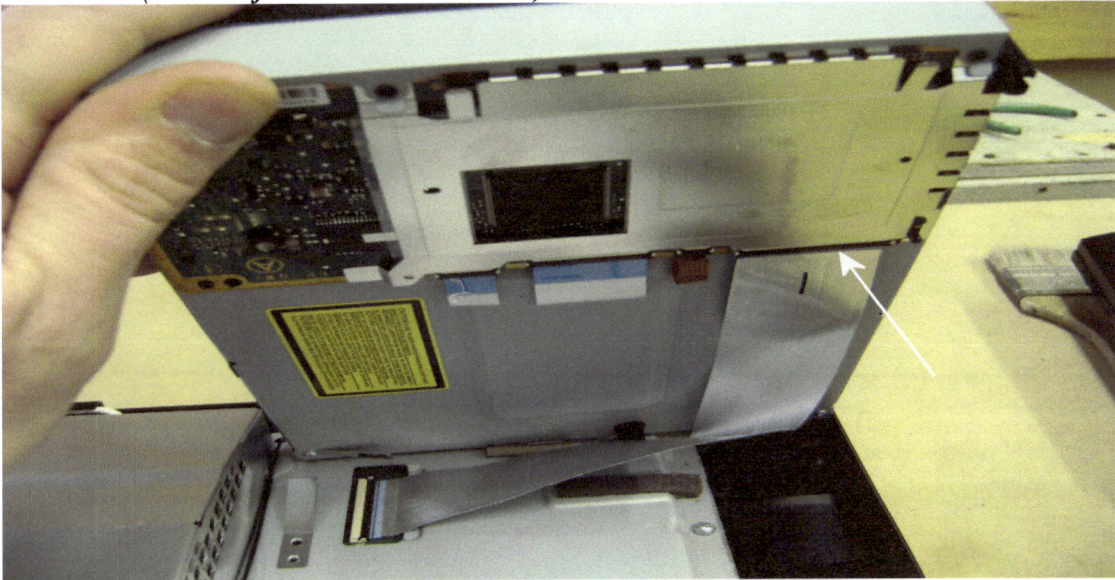

- Wi-Fi / Bluetooth Board Removal

The main board is located in front of the power supply. The secondary board connects to the main one through a black cable and is located at the back of the system behind the disc drive close to the AV connections. Use Pictures 16 and 17 to identify and remove these boards.

Picture 16 (Main Wi-Fi board removal)

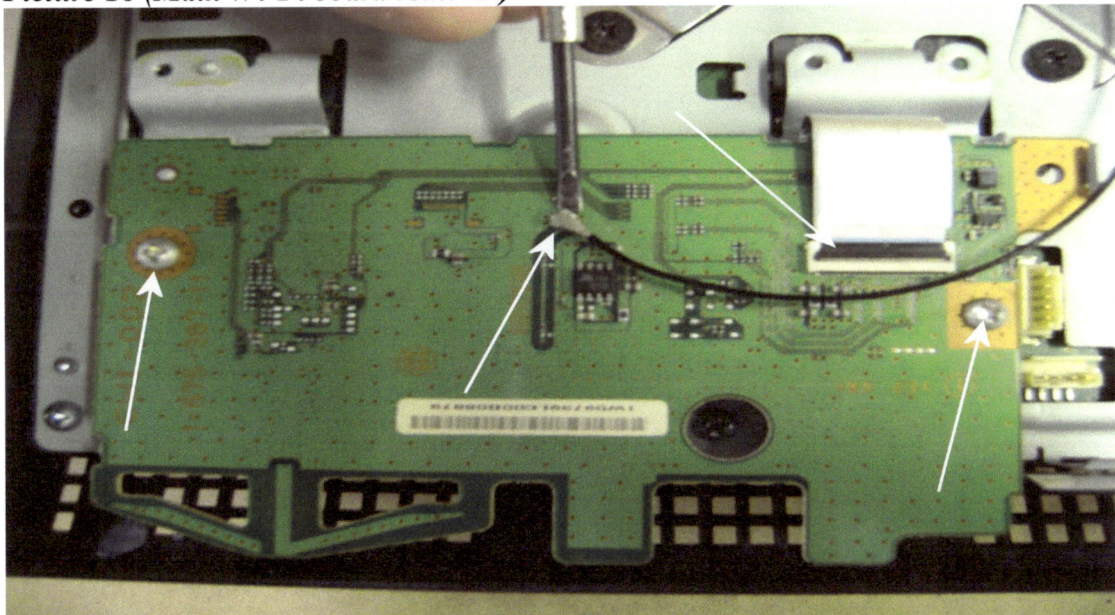

Picture 17 (Secondary Wi-Fi board removal)

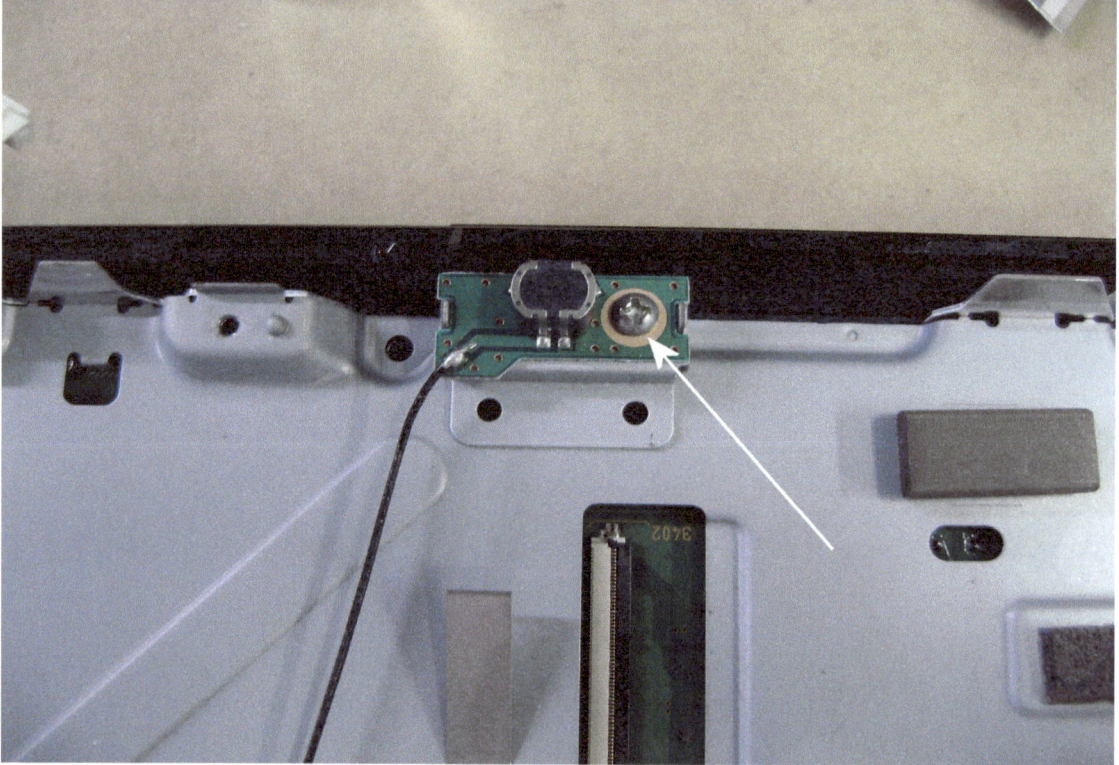

- Motherboard Removal

There are several screws now to remove so the motherboard can come out of the plastic casing and metal cover. The screws to remove are shown in Picture 18. Note that there are arrows pointing to the screws to remove on the system itself also. Disconnect the cables that are on the motherboard for now. Remove the hard drive door on the left side of the console. Refer to Picture 2 on page 7 to identify it if necessary. Remove the large blue screw and pull the hard drive out by the handle. The hard drive will come out by pulling to the right and then out or just come straight out. The motherboard now can be lifted up out of the plastic bottom of the console like in Picture 19. The plastic cover over the AV jacks on the back of the motherboard will pull off as shown in Picture 20. Pull the top metal cover off of the motherboard as shown in Picture 21. Flip the motherboard over and disconnect the fan cable from the board as in Picture 22. Separate the motherboard from the heat sink / fan assembly as shown in Picture 23. Now the system has been completely disassembled down to the motherboard for repair. Please refer to your diagnosis chapter for repair procedures.

Picture 18 (Identification of screws for motherboard removal)

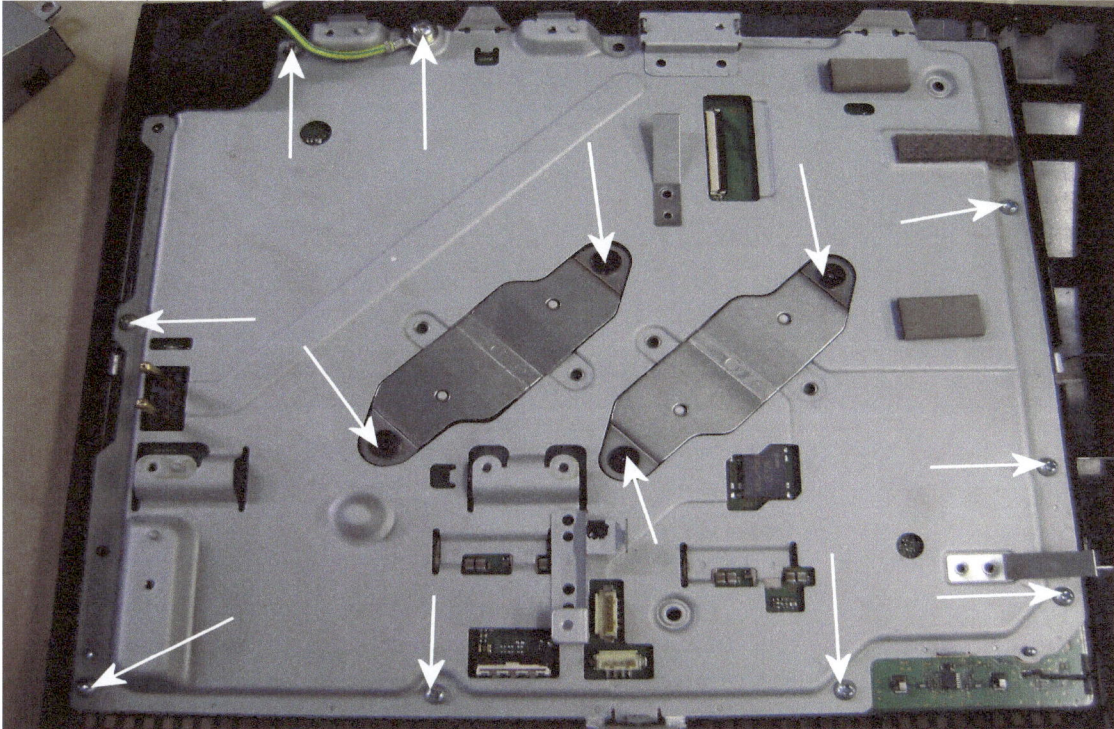

Picture 19 (Motherboard removal from bottom console casing)

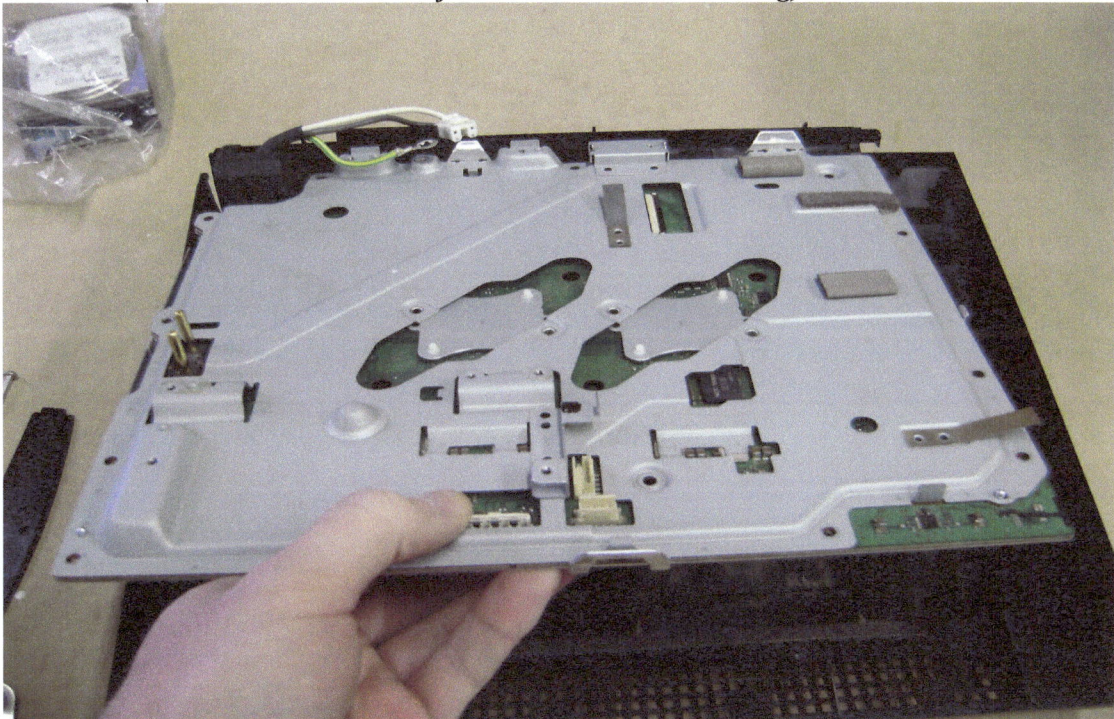

Picture 20 (Removal of AV cover)

Picture 21 (Removal of top metal cover to motherboard)

Picture 22 (Disconnecting fan cable from motherboard)

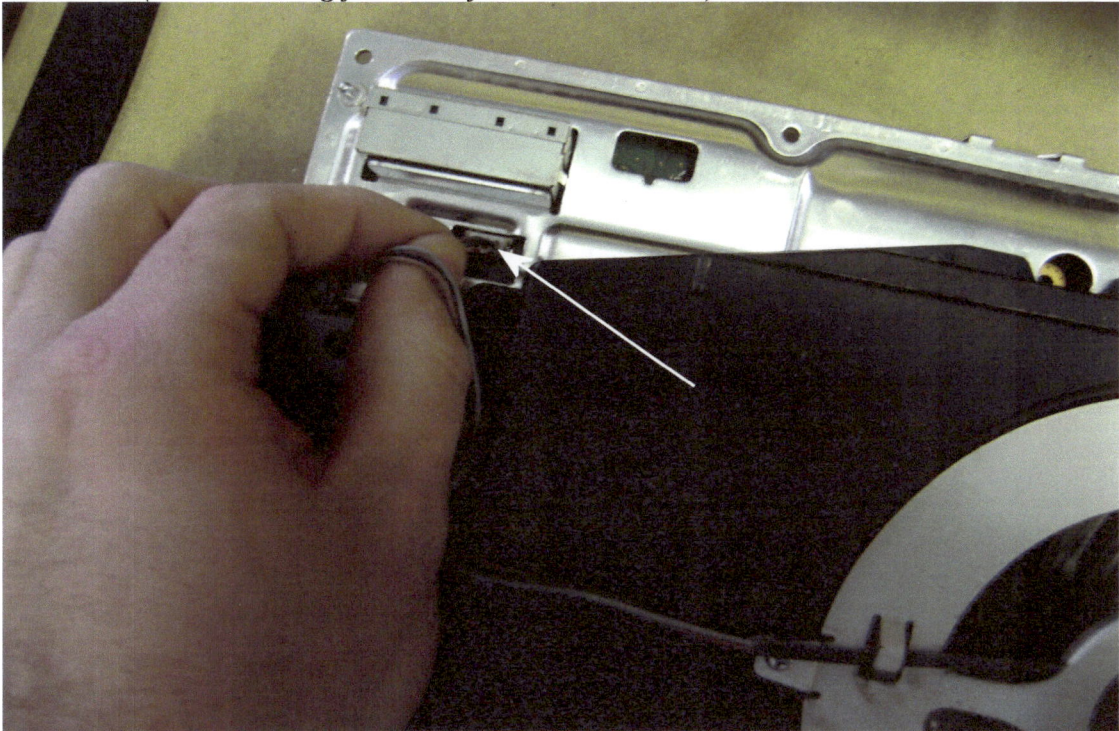

Picture 23 (Separating motherboard from heat sink/fan assembly)

Chapter 4 – Disc Drive Issues

There are several different symptoms that are caused by a defective disc drive. First, we need to disassemble the system and remove the disc drive. View Chapter 3 for details on this. Next, identify which drive is in your system. There are 2 distinct styles of disc drives and they are not interchangeable. Use Picture 24 to help identify the drive. The two different drives have been labeled "A" and "B", which we will use to reference them from now on. At this point, you should know that further disassembly and repair of the drive will be of a high difficulty level. It will take time and patience!

Please note that it would be much easier to purchase a replacement drive. Many websites sell the complete disc drive. Beware of purchasing a generic or aftermarket part. Make sure the replacement drive looks just like the original. Confirm pictures of the product being purchased and even contact the seller's customer service for assistance to ensure you receive the correct product. Make sure to keep the circuit board that was removed from the broken drive as you will need this for re-installation of the replacement! If you are up to the challenge and want to save a little money, proceed to the corresponding symptom you are experiencing with your drive below. The rest of this chapter is broken into 2 separate sections for drive type "A" on the next page and type "B" on page 32.

Picture 24 (Disc drive types)

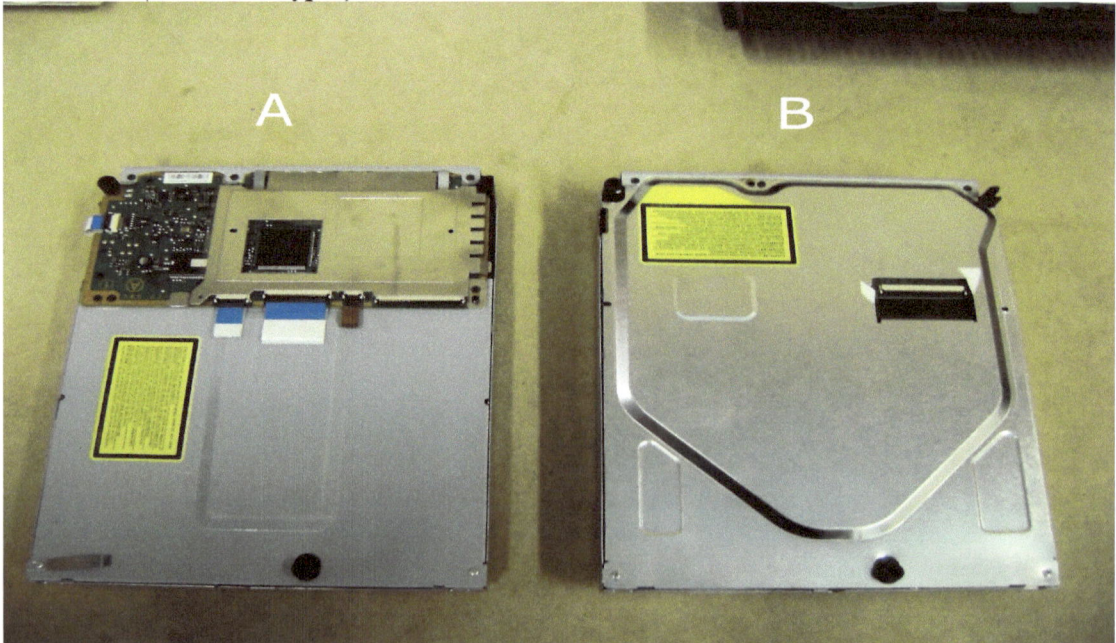

Disc Drive Type "A"

Symptom: Drive will not eject a disc.

This is caused by alignment problems in gear system. For this repair follow the "Drive Disassembly" and "Gear System Realignment" to fix these problems.

Symptom: Drive will not accept a disc.

Refer to the solution for the symptom above, "Drive will not eject disc" as the repair procedure is identical.

Symptom: Drive will not read a disc.

This can be caused by either gear alignment problems or a defective laser. Visit the "Further Diagnosis of Disc Reading Issues" section to identify what part of the drive is causing this problem.

Symptom: Video freezes and often is followed by the system not responding.

This is caused by a defective laser. View the sections "Drive Disassembly" and "Laser Lens Replacement" to fix this issue.

Drive Disassembly

Flip the drive upside down and remove the 7 black screws. There will be 4 ribbon cables to disconnect. Use Picture 25 for reference. Be careful and caution with the next step. Gently flip the circuit board over to the left to reveal the bottom of the circuit board. Disconnect the last cable as shown in Picture 26. It may be helpful to use a set of tweezers to get this small cable out. Pull the 2 metal covers on the top and bottom of the drive off as shown in Picture 27. Now lay the disc drive top up on a flat surface. As shown in Picture 28, remove 5 small black screws and the white magnetic disc holder pointed to. Remove the top of the drive and sit it aside. Remove any disc that may be inside. Clean out the inside of the drive very thoroughly (including the laser lens). For cleaning the lens, we recommend using cotton swabs and a lens cleaning solution. Make sure not to use too much solution and dry the lens before reassembly. If the issue was the drive not reading discs, you may want to reassemble it for testing before going any further with repairs.

Picture 25 (Screw removal for drive type A)

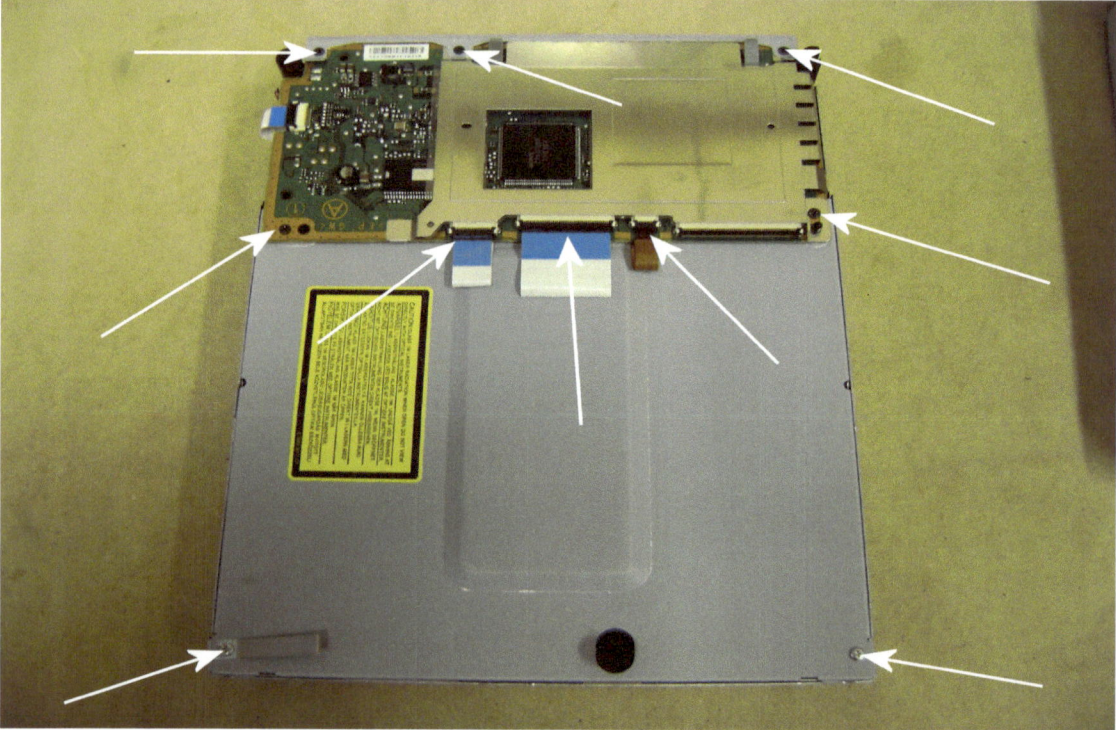

Picture 26 (Drive type A small cable removal)

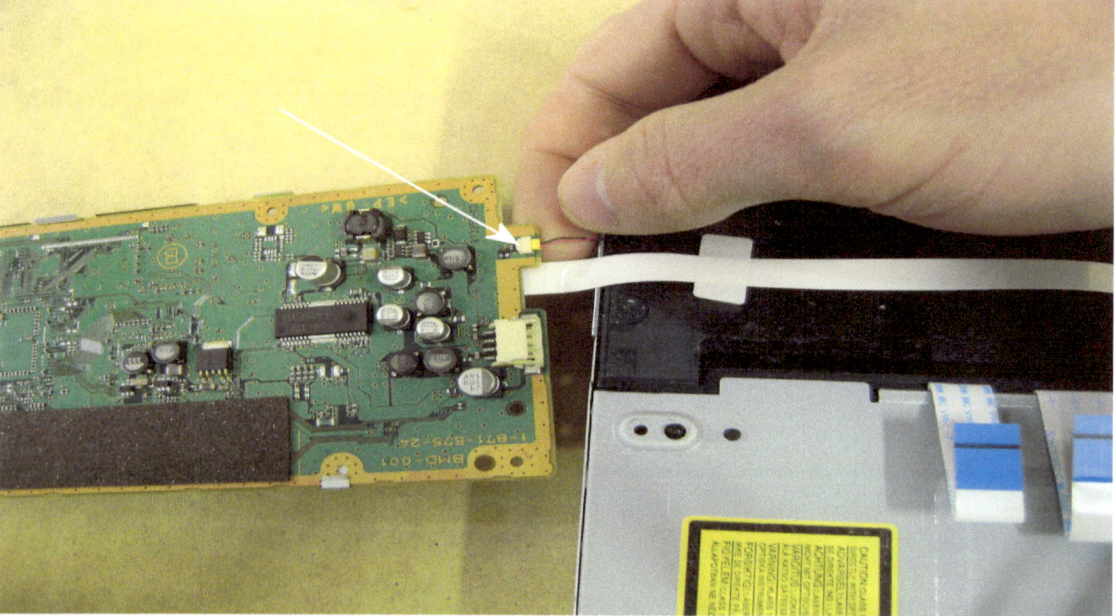

Picture 27 (Drive type A metal cover removal)

Picture 28 (Drive type A top removal)

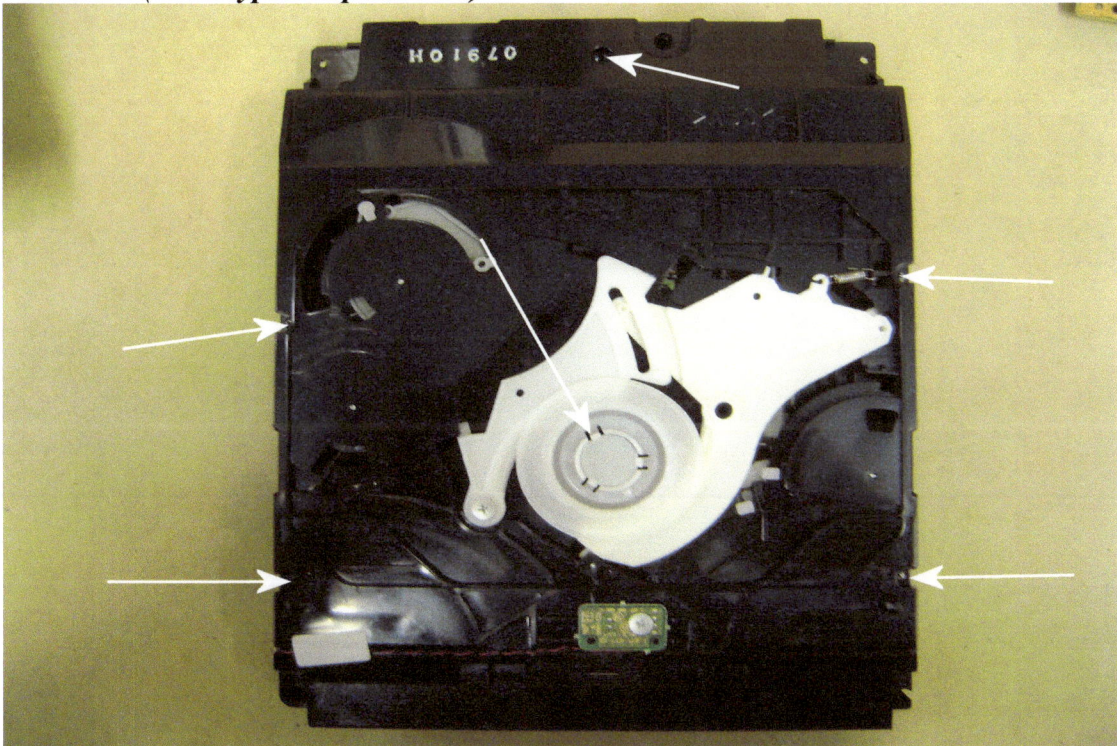

Gear System Realignment

Remove the screw pointed to in Picture 29 and pull the gear system off. It should automatically retract back in like Picture 30. If it does not, turn it over and realign the gears so it retracts all the way in while moving freely back and forth. Use caution when handling the gear assembly. This gear system is very fragile. If any gears are missing or broken at all on any part of the drive, you will unfortunately have to buy a complete replacement drive.

Picture 29 (Drive type A gear removal)

Picture 30 (Drive type A gear assembly)

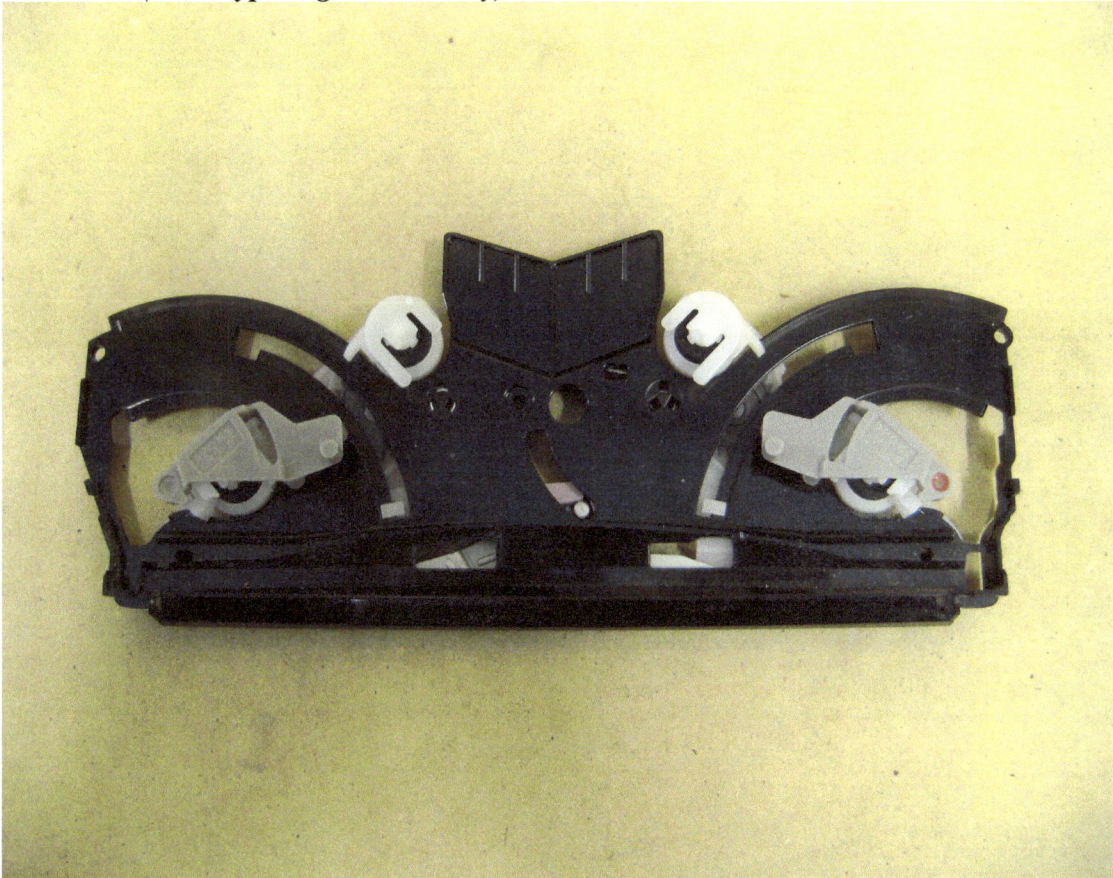

     Now that the gear assembly fixed, move on to the next step. There are two large plastic black bars on the left and right side of the drive that move back and forth. Both are pointed to in Picture 31. Pull the left one up until it is no longer touching the gear system at the bottom of the drive pointed to in Picture 31. You will have to apply some force for this. The black bar on the right side should be down as far as it can go. Very closely examine Picture 31 and make sure your drive looks exactly the same after repositioning the bars on the right and left side including the gap pointed to in the upper right side. If everything is moved into the correct position, move on to the "Drive Reassembly and Testing" section. If the bars on each side of the drive will not move into the position, move on to the next step.

Picture 31 (Drive type A gear system correctly configured)

Flip the drive upside down and remove the screw pointed to in Picture 32. Remove the metal bar that was secured by the screw just removed. Note the metal tabs sticking out on each end. There are holes in the bottom of the black bars you just aligned that these metal tabs should be inserted into. These holes are shown in Picture 33. Reinstall the metal bar ensuring the ends are placed into their respective hole on each plastic side bar. Secure it in place with the previously used screw. Double check again that the ends of the metal bar are in the holes as this is an important step. Now you should be able to move these bars into their correct position. Realigning the gear system is very tricky and sometimes takes several attempts before success. After this, view the "Drive Reassembly and Testing" section of the chapter to complete the repair.

Picture 32 (Drive metal bar realignment step 1)

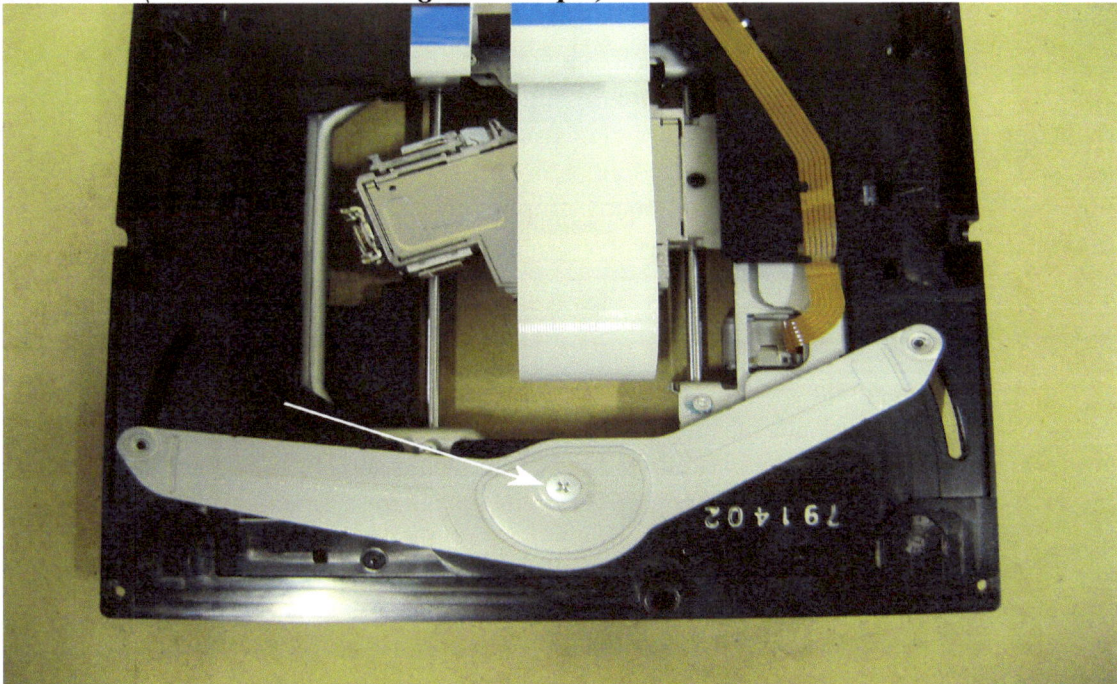

Picture 33 (Drive metal bar realignment part 2)

Further Diagnosis of Disc Reading Issues

The metal cover needs to be removed from the drive so you can see how it is working. Follow the "Drive Disassembly" section on page 21, making sure to stop after removing the metal covers and circuit board from the drive. Reinstall the circuit board that was just removed, ensuring to reconnect all 5 of the cables. Use a couple of the previously used screws to secure the board temporarily in place. Install the drive back into the system connecting the ribbon cable on the bottom and the other cable on the left side. Reconnect the power cable and turn the system on. Insert a disc into the drive and watch what it is doing.

Does it accept the disc all the way into the drive and start spinning? If it does not accept the disc in, revisit the "Gear System Alignment" section again. If the drive accepts the disc and spins up without reading it that means the laser is defective. Follow the section on "Laser Lens Replacement" to fix this problem. If the disc spins up and stops shortly after or does not keep a consistent speed, the laser assembly is defective. View the section on "Complete Laser Assembly Replacement" to repair this issue. If the disc does not spin, remove the white magnetic disc and observe the black spindle head in the drive. Is it inside of the center of the disc like in Picture 34? Does the disc spin freely if it is moved? If your answer was yes to these questions, the laser is defective and the "Complete Laser Assembly Replacement" section will fix the problem. If your answer was no to these questions, the issue is the gear system alignment not feeding the disc correctly. View the "Gear System Realignment" section for further repair.

Picture 34 (Disc correctly inserted inside drive spindle)

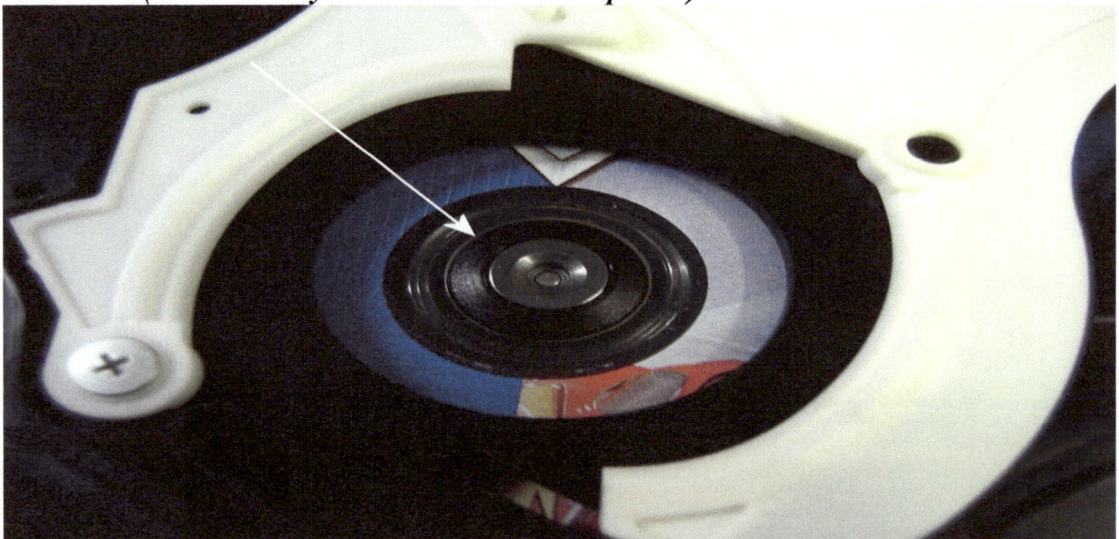

Laser Lens Replacement

To replace the laser lens, there are 4 small black screws and 1 ribbon cable to remove as shown in Picture 35. Remove the 4 metal tabs that were held down by the screws just removed. Now move the left and right bars out of the assembly. Slide the laser off the bars. The laser is sensitive to static electricity so make sure to ground yourself before touching it (touch metal first or wear an ESD strap). There is a part number on a white sticker located somewhere on the laser lens as shown in Picture 36. It probably is KES-400A (1 laser eye) or KES-410A (2 laser eyes). Use this part number to purchase a replacement. Many websites sell this laser assembly. Beware of purchasing a generic or aftermarket laser lens. Make sure what you are purchasing is an original part. Remove the black screw on the bottom that will release the white gear system attached to the laser like in Picture 37. Install the new laser in reverse order of disassembly. Reinstall the white gear assembly to the lens with the screw previously used. Slide the bars back into the replacement laser lens. Place the bars in the previous position. Secure them in place with the previously used metal tabs and screws. Reconnect the ribbon cable to the laser lens.

Picture 35 (Screws for drive type A laser lens removal)

Picture 36 (Drive type A laser lens identification)

Picture 37 (Drive type A gear removal from laser lens)

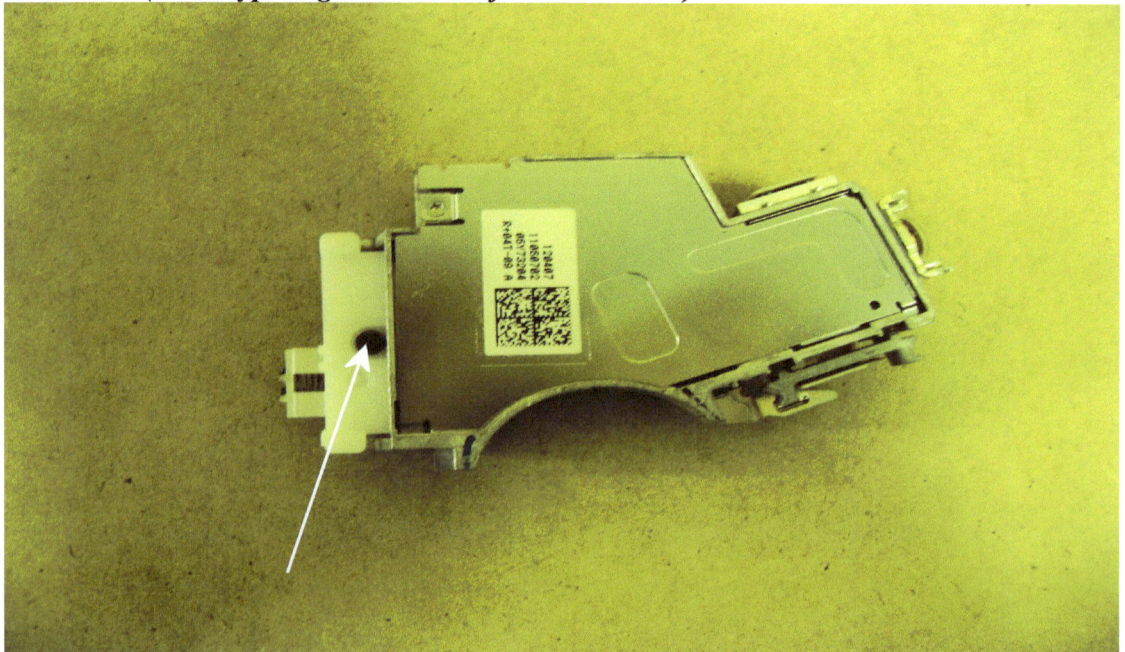

Complete Laser Assembly Replacement

First, identify the part number of the laser assembly you have. It is located on a white sticker and engraved on the metal deck as shown in Picture 38. Chances are the part number is KEM-400AAA. If not, it is the KEM-410AAA. These two parts are not interchangeable, so verify which one you have. Use this part number to order a replacement laser assembly. Many websites sell this laser assembly. Beware of purchasing a generic or aftermarket part. Make sure what you are purchasing is an original part. There are 4 screws to remove and 2 ribbon cables to disconnect. Use picture 39 for reference if needed. Lift the laser deck up and out. There are plastic or rubber spacers underneath each screw that need to be saved for re-installation. Once you have a replacement laser deck, reinstall by seating it back in place remembering to use the spacers underneath each screw. Connect the 2 ribbon cables previously disconnected. Follow the "Drive Reassembly and Testing" section to finish the repair.

Picture 38 (Drive type A laser assembly identification)

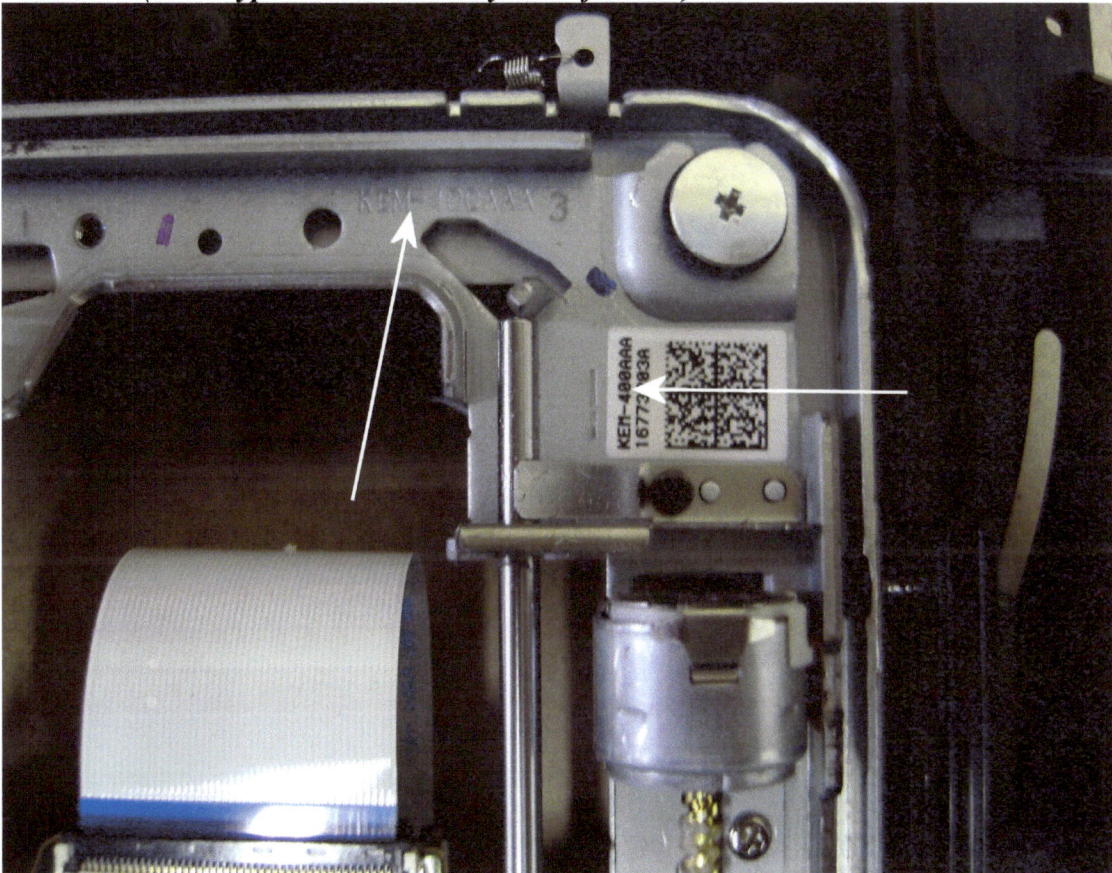

Picture 39 (Drive type A laser deck removal)

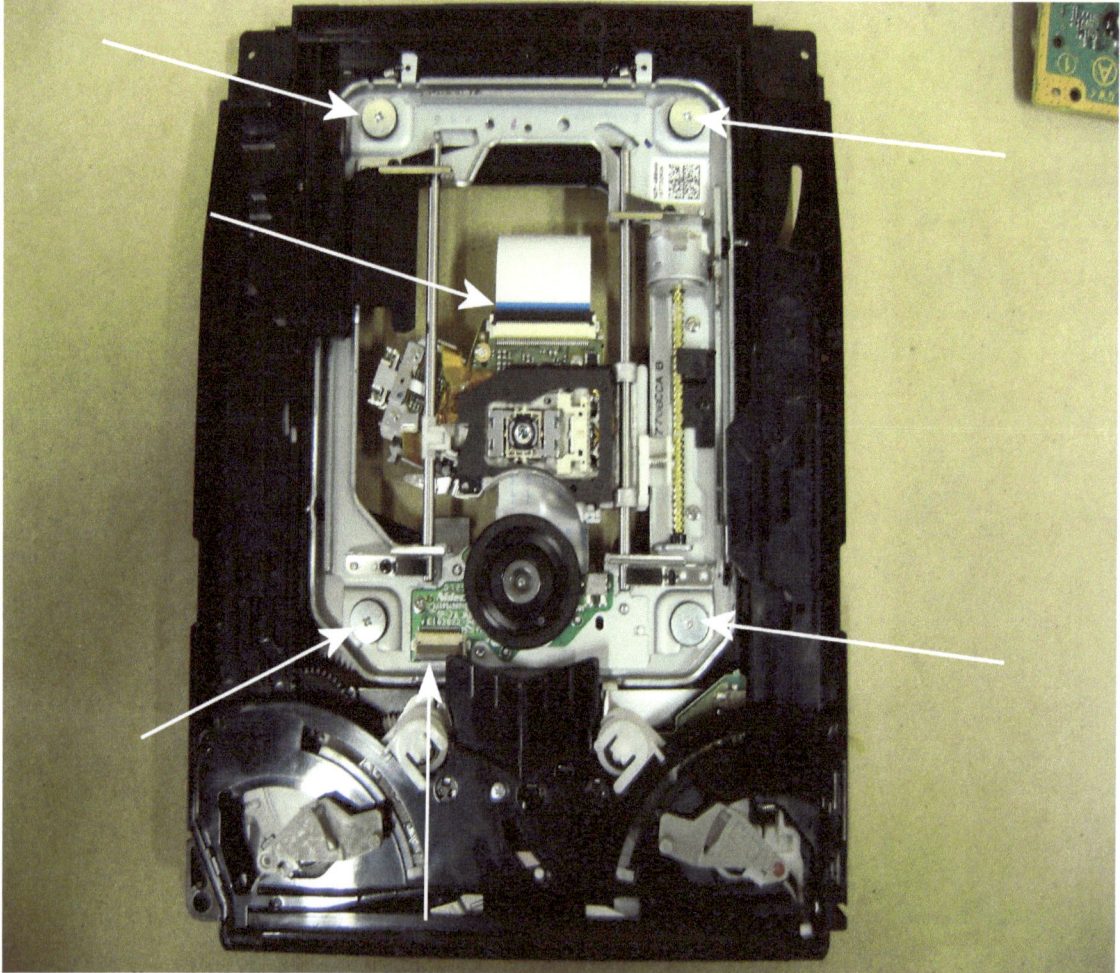

Drive Reassembly and Testing

If previously removed, reinstall the gear system using the corresponding screw. Put the top back on the drive and at this point compare yours to Picture 40. If it looks like the incorrect one, remove the top again and move the white gear down. Put the top back on and make sure it does not move back up blocking that hole pointed to in the picture. Secure the top of the drive in place with the 5 previously used black screws. Turn the drive over and reinstall the circuit board if previously removed. Reconnect all 5 cables and use a couple screws to secure the board in place temporarily. Test it for proper operation before completely reassembling the drive and system. This is done to save time as occasionally it takes several repair attempts before the drive works properly.

Put the white magnetic disc holder back in the top center of the drive and reconnect the 2 cables from the system to the drive. Power the system on and insert a disc. If the drive does not work properly, first double check all cable connections. If these are correct, revisit the repair sections again until it works correctly. It does take some time and patience! When the drive is working correctly, disconnect it from the system. Disconnect the circuit board from the bottom of the drive again. Place it back inside of the metal covers and seat the circuit board back on. Secure all 7 screws and reconnect the 5 cables. Now place the drive back in the system connecting the 2 cables. Use Chapter 12 for help reassembling the system.

Picture 40 (Top gear assembly of drive)

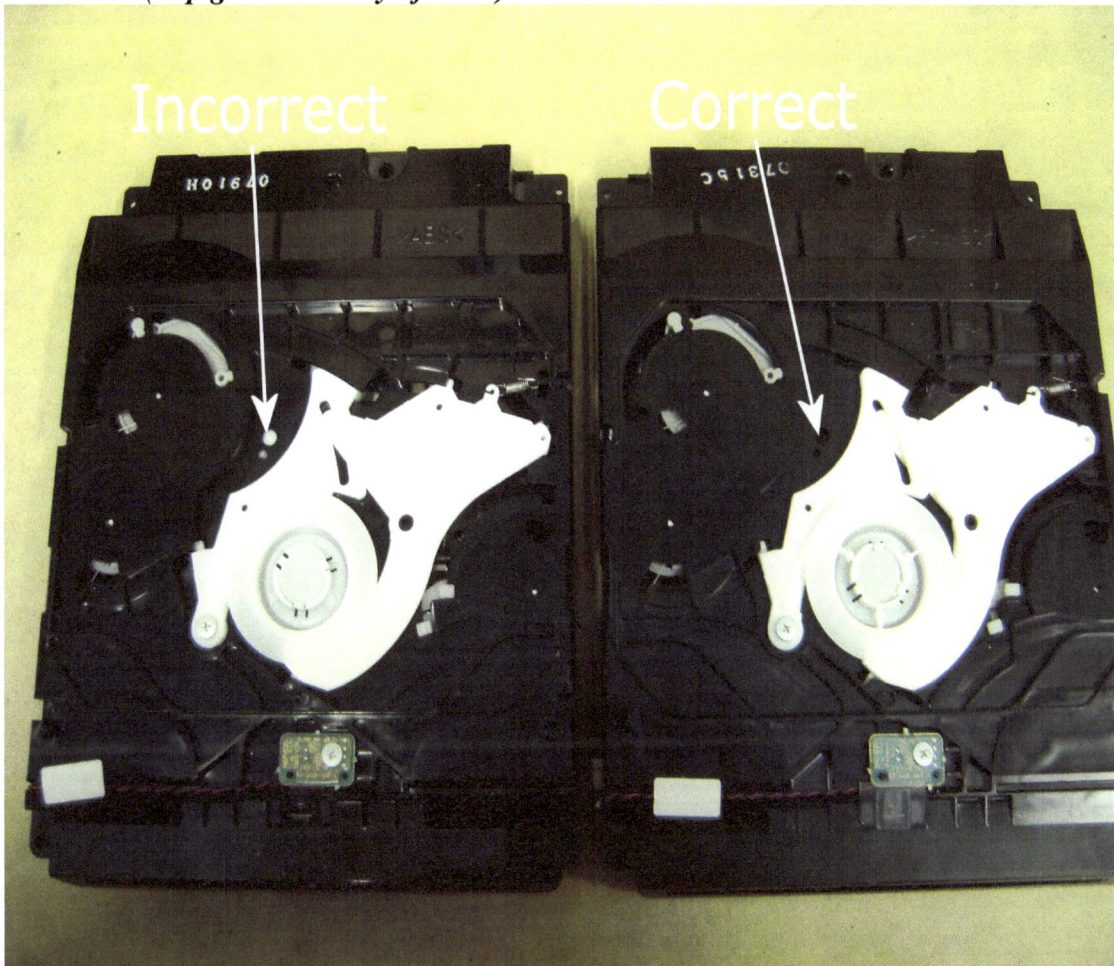

Disc Drive Type "B"

Symptom: Drive will not eject a disc.

This is caused by alignment problems in gear system. For this repair follow the "Drive Disassembly" and "Gear System Realignment" to fix these problems.

Symptom: Drive will not accept a disc.

Refer to the solution for the symptom above, "Drive will not eject disc" as the repair procedure is identical.

Symptom: Drive will not read a disc.

This can be caused by either gear alignment problems or a defective laser. Visit the "Further Diagnosis of Disc Reading Issues" section to identify what part of the drive is causing this problem.

Symptom: Video freezes and often is followed by the system not responding.

This is caused by a defective laser. View the sections "Drive Disassembly" and "Laser Lens Replacement" to fix the issue.

Drive Disassembly

Turn the drive upside down and remove the 5 screws. Use Picture 41 for reference. Pull both metal covers off of the drive. There are 2 more screws to remove and 3 ribbon cables to disconnect as shown in Picture 42. Remove the circuit board and pull the top metal cover off of the drive. Lay the drive down with the top facing up. As shown in Picture 43, remove 2 small black screws and the white magnetic disc holder pointed to. Lift the tabs on the left and right side of the drive to remove the top. Remove any disc that may be inside. Clean out the inside of the drive very thoroughly (including the laser lens). For cleaning the lens, we recommend using cotton swabs and a lens cleaning solution. Make sure not to use too much solution and dry the lens before reassembly. If the issue was the drive not reading discs, you may want to reassemble it for testing before going any further with repairs.

Picture 41 (Screw removal for drive type B)

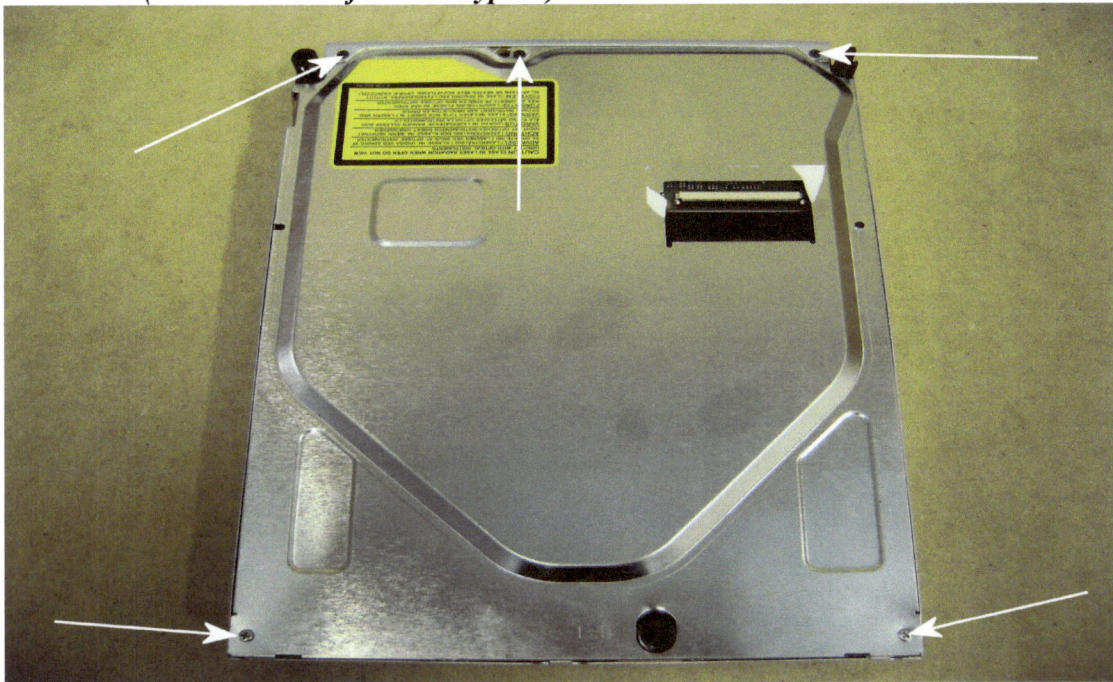

Picture 42 (Drive type B circuit board removal)

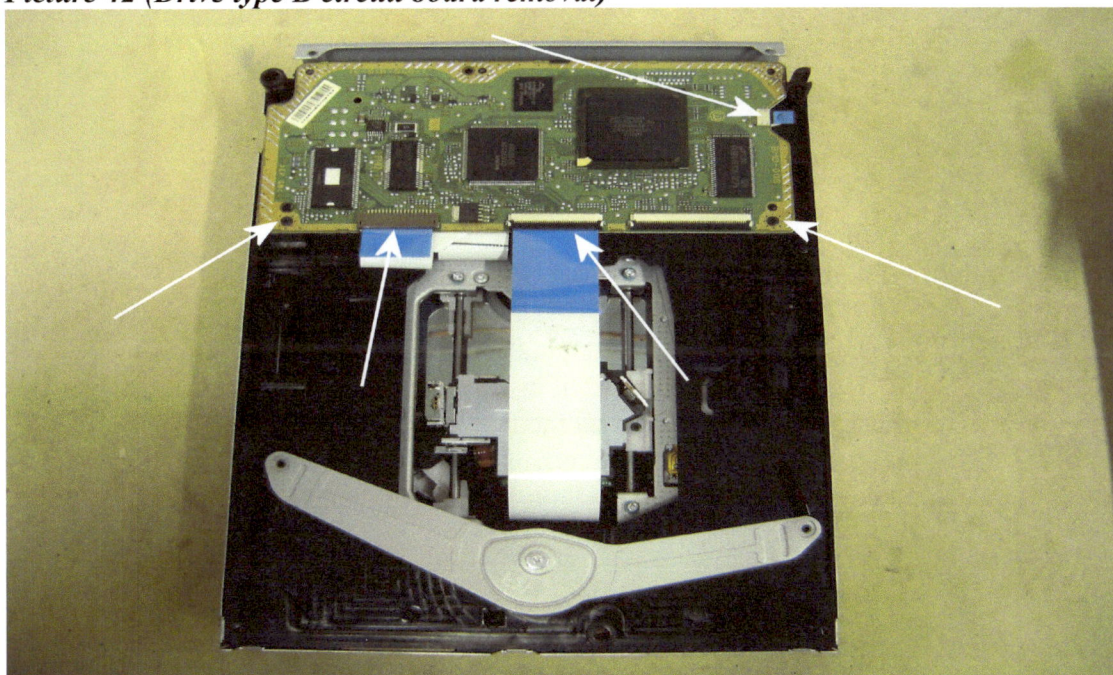

Picture 43 (Drive type B top removal)

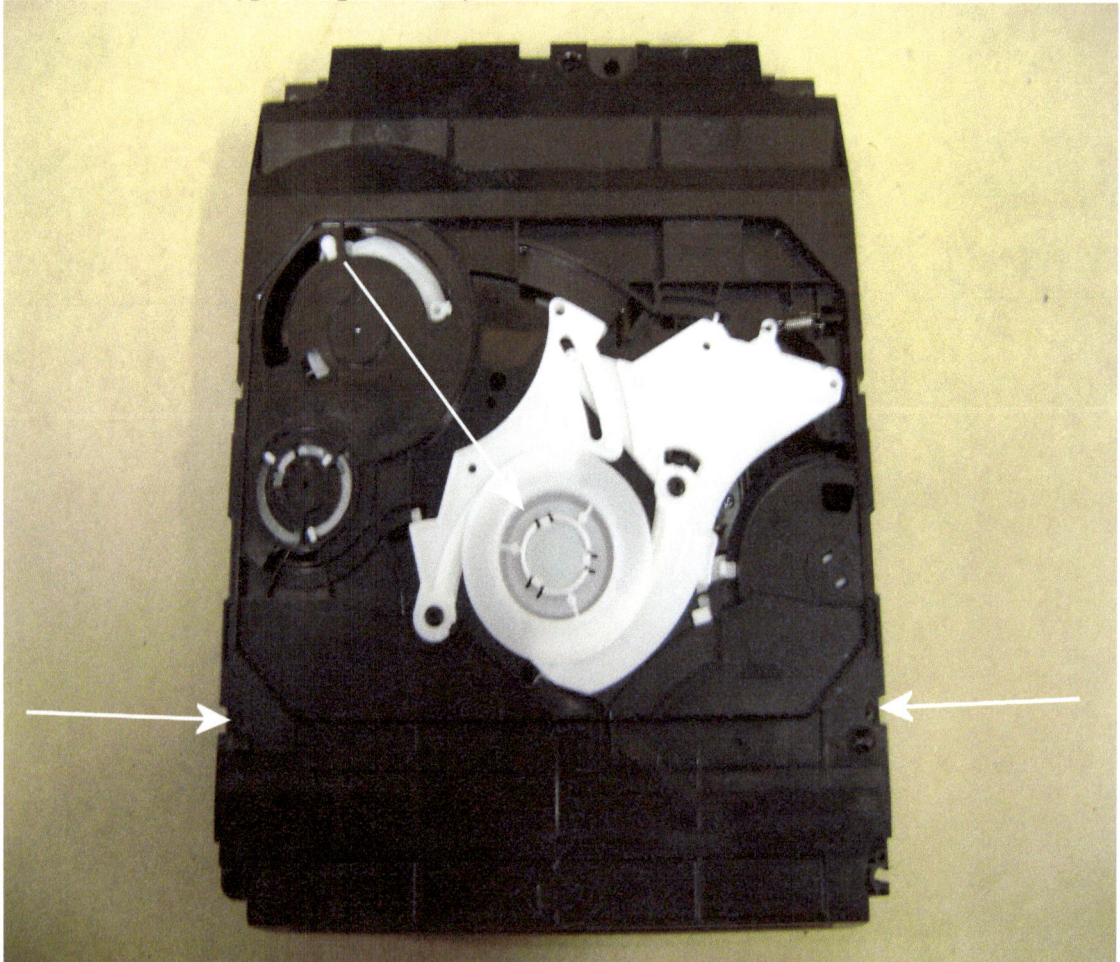

Gear System Realignment

Pull off the gear system by lifting up on the areas pointed to in Picture 44. The gear system should automatically retract back in like Picture 45. If it does not, turn it over and realign the gears until it retracts all the way in while moving freely back and forth. Use caution when handling the gear assembly as to not accidentally pull any gears out. If any gears are missing or broken at all on any part of the drive, you will unfortunately have to buy a replacement drive.

Picture 44 (Drive type B gear system removal)

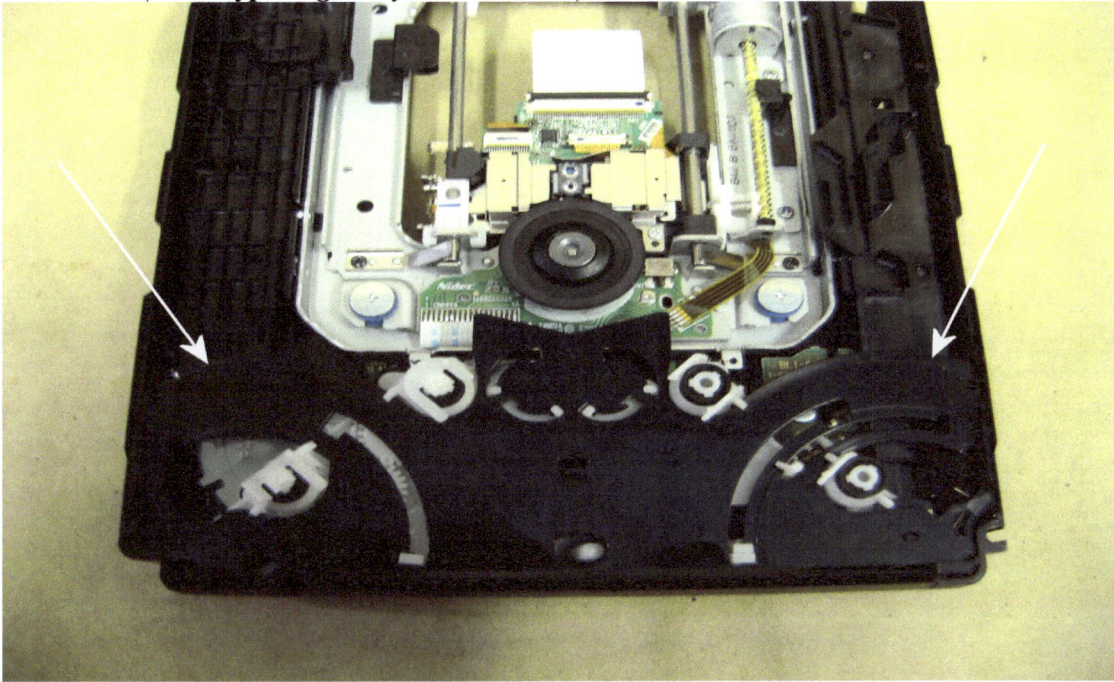

Picture 45 (Drive type B gear system)

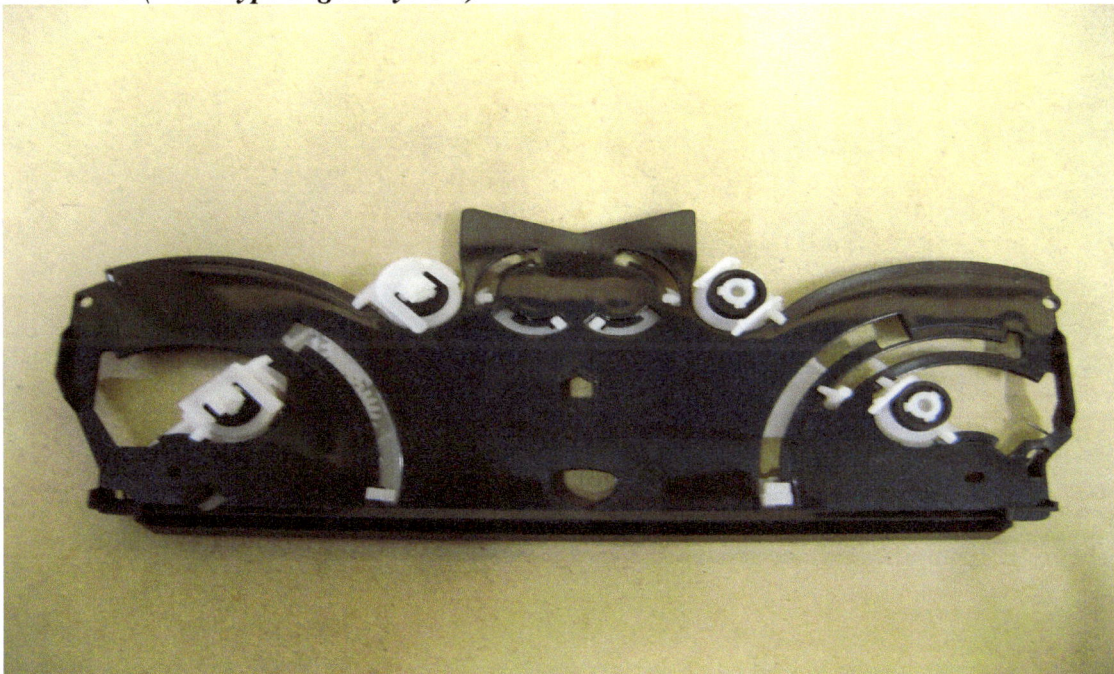

There are two plastic black bars on the left and right side of the drive that move back and forth. Both are pointed to in Picture 46. Pull the left one up until it is not touching the gear system at the bottom of the drive also pointed to in Picture 46. Force will have to be applied for this. The black bar on the right side should be down as far as it can go. Very closely examine Picture 46 and make sure your drive looks exactly the same after repositioning the bars on the right and left side including the gap pointed to in the upper right side. If everything is moved into the correct position, move on to the "Drive Reassembly and Testing" section. If the bars on each side of the drive will not move into position, move on to the next step.

Picture 46 (Drive type B gear system correctly configured)

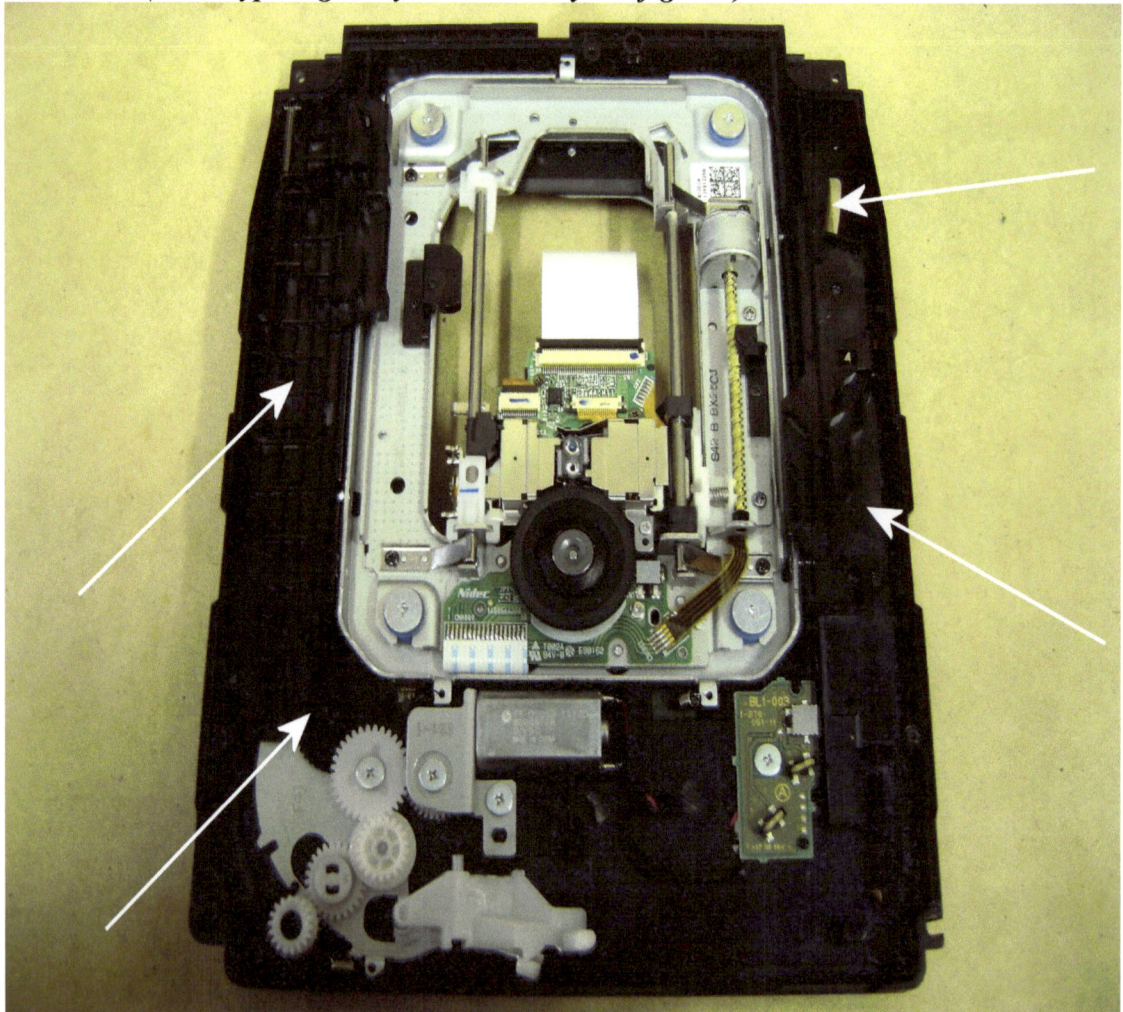

Flip the drive upside down and remove the screw pointed to in Picture 47. Remove the metal bar that was secured by the screw just removed. Note the metal tabs sticking out on each end. There are holes in the bottom of the black bars you just aligned that these metal tabs should be inserted into. These holes are shown in Picture 48. Reinstall the metal bar ensuring the ends are placed into their respective hole on each plastic side bar. Secure it in place with the previously used screw. Double check again that the ends of the metal bar are in the holes as this is an important step. Now the bars should move into the correct positions. Realigning the gear system is very tricky and sometimes takes several attempts before success. After this, view the "Drive Reassembly and Testing" section of the chapter to complete the repair.

Picture 47 (Drive metal bar realignment step 1)

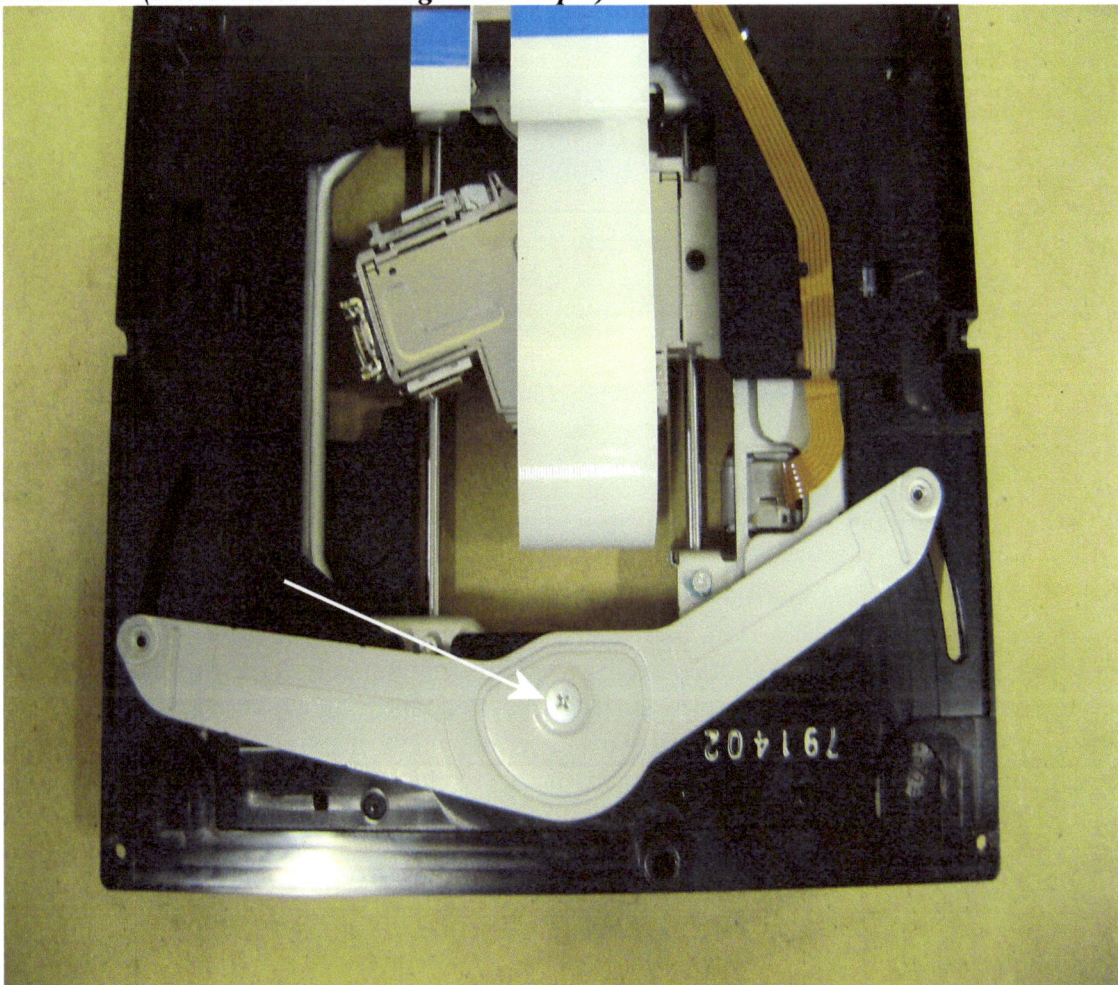

Picture 48 (Drive metal bar realignment part 2)

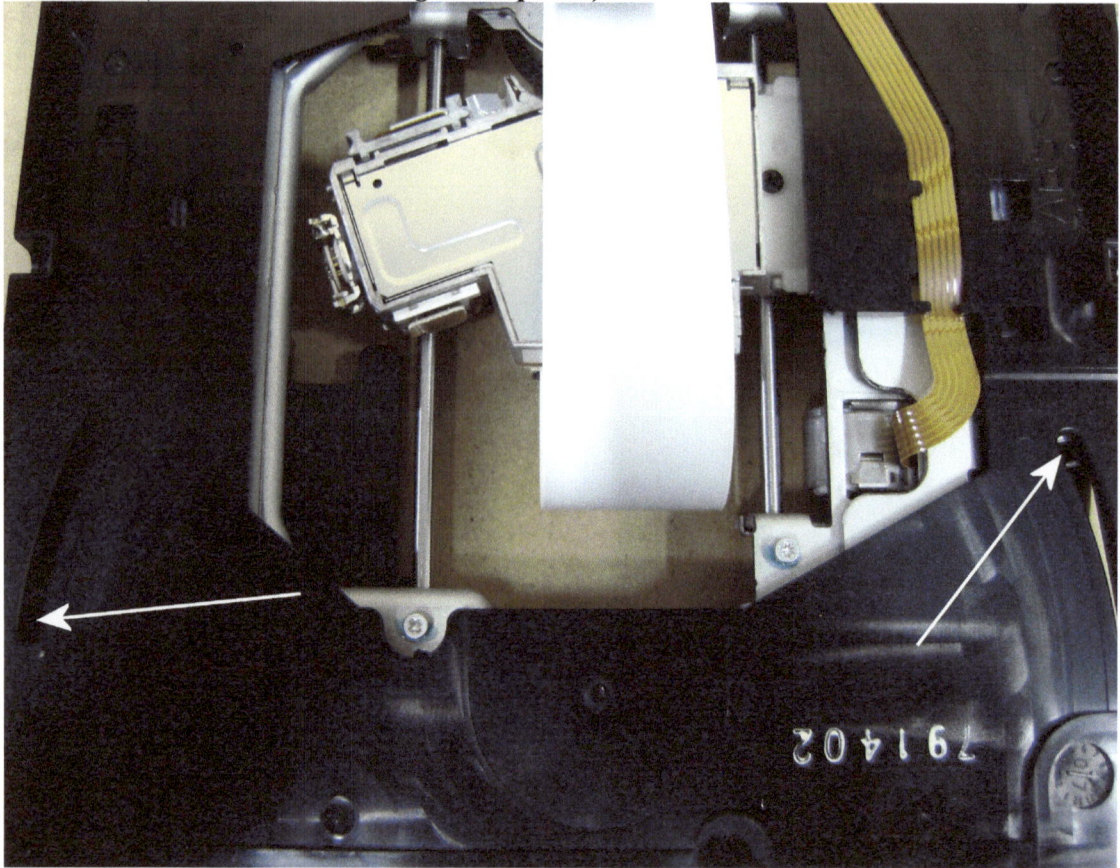

Further Diagnosis of Disc Reading Issues

The metal cover needs to be removed from the drive so you can see how it is working. Refer to the "Drive Disassembly" section on page 33 to remove the metal covers only. After this, install the drive back into the system connecting the ribbon cable on the bottom and the other cable on the left side. Reconnect the power cable and turn the system on. Insert a disc into the drive and watch what it is doing. Does it accept the disc in and start spinning? The white magnetic disc holder will be moving with the disc if it is spinning. If it does not accept the disc in, revisit the "Gear System Alignment" section again. If the drive accepts the disc and spins up without reading it that means the laser is defective. Follow the section on "Laser Lens Replacement" to fix this problem. If the disc spins up and stops shortly after or does not keep a consistent speed, use the section "Complete Laser Assembly Replacement" to repair this issue. If the disc does not spin, remove the white magnetic disc and observe the black spindle head in the drive. Is it inside of the center of the disc like in Picture 49? Does the disc spin freely when moved? If your answer was yes to these questions, the laser is defective and the "Complete Laser Assembly Replacement" section will fix the problem. If your answer was no to these questions, the issue is the gear system alignment not feeding the disc correctly. View the "Gear System Realignment" section for further repair.

Picture 49 (Disc correctly inserted inside drive spindle)

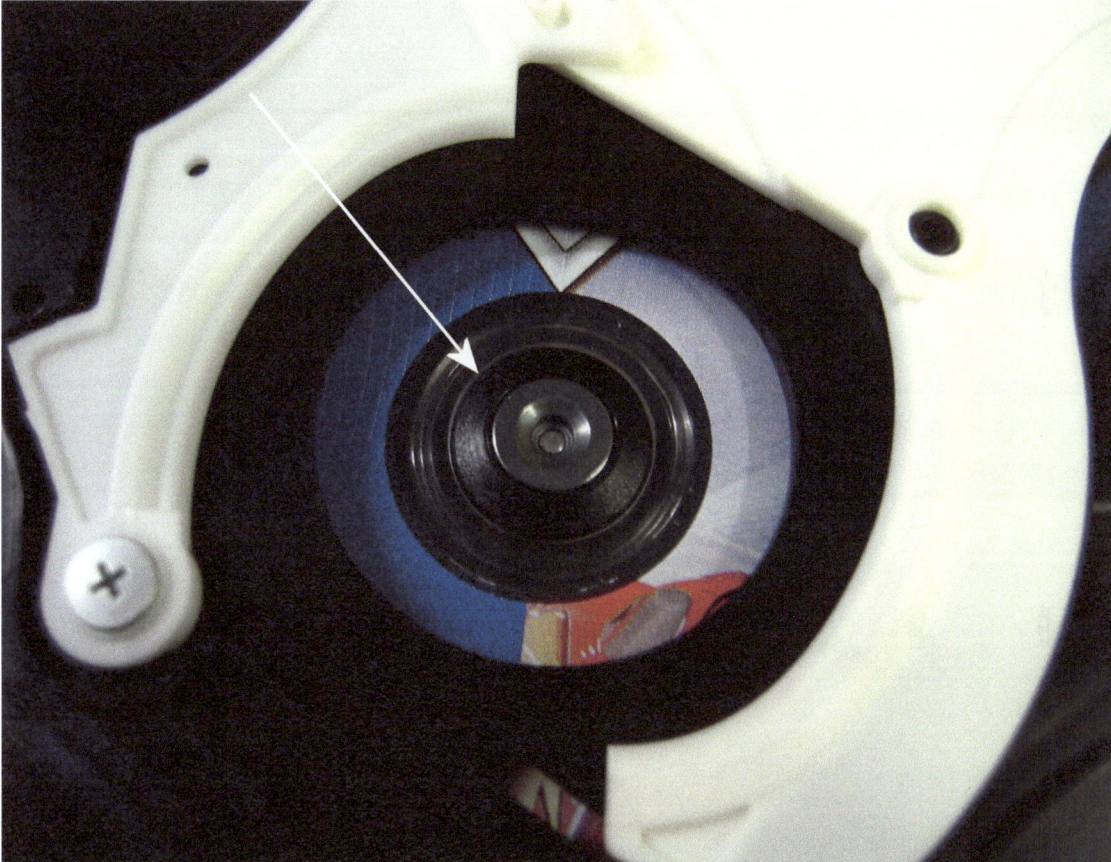

Laser Lens Replacement

To replace the laser lens, there are 4 small black screws and 1 ribbon cable to remove as shown in Picture 50. Remove the 4 metal tabs that were held down by the screws just removed. Now slide the right bar out of the assembly. The laser can now slide off of that bar. The laser is sensitive to static electricity so make sure to ground yourself before touching it (touch metal first or wear an ESD strap). The laser should have 2 lenses on it like in Picture 51. If so, the part number is KES-410A or KES-

410ACA. If the laser has only 1 lens, the part number is KES-400A. Use this part number to purchase a replacement. Many websites sell this laser assembly. Beware of purchasing a generic or aftermarket part. Make sure what you are purchasing is an original part. Remove the black screw on the bottom that will release the white gear system attached to the laser like in Picture 52. Install the new laser in reverse order of disassembly. Reinstall the white gear assembly to the lens with the screw previously used. Slide the bar back into the replacement laser lens. Align the left bar with the space on the lens while pushing the right bar back in place. Secure them in place with the previously used metal tabs and screws. Reconnect the ribbon cable to the laser lens.

Picture 50 (Drive type B laser lens removal)

Picture 51 (Drive type B laser lens identification)

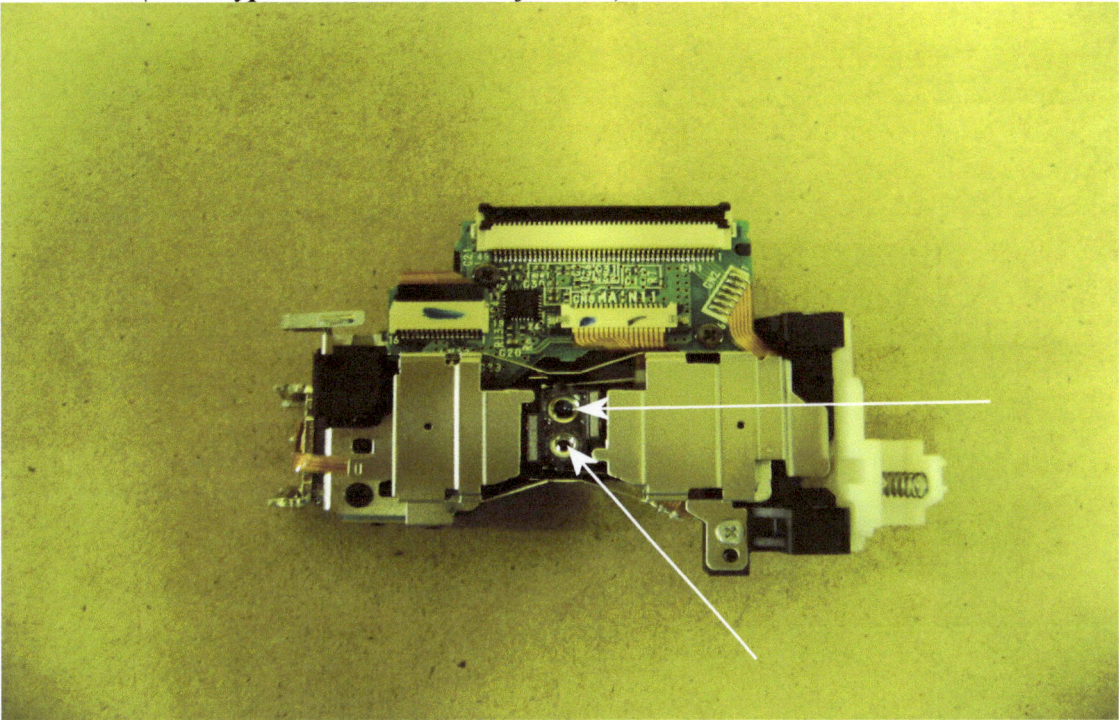

Picture 52 (Drive type B laser lens gear removal)

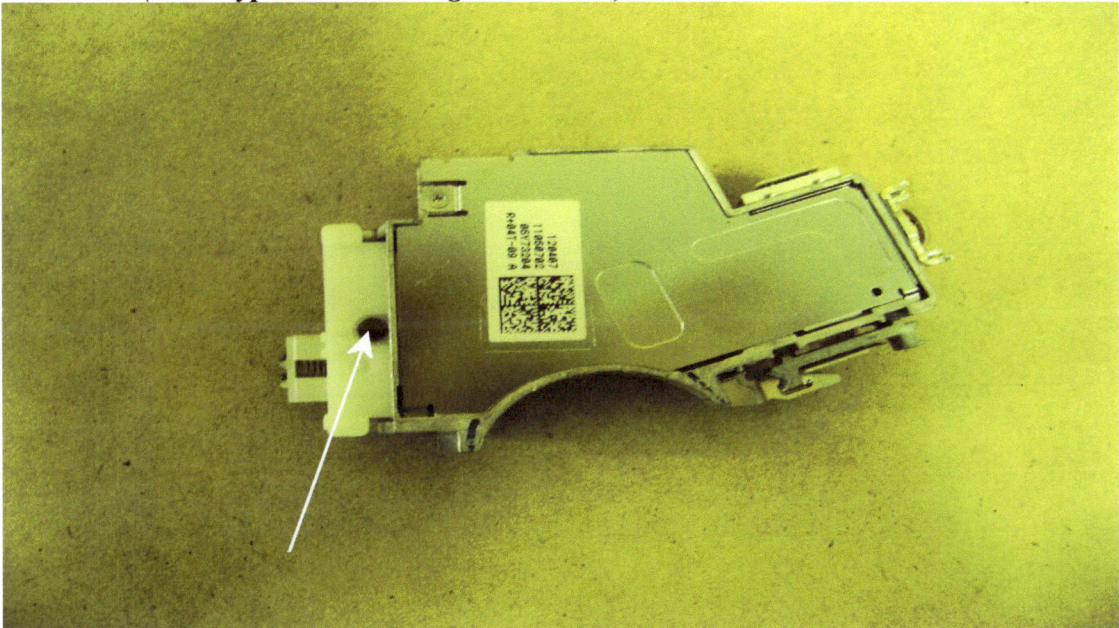

Complete Laser Assembly Replacement

First, identify the part number of the laser assembly you have. It is engraved on the metal deck as shown in Picture 53. The part number should be KEM-410ACA, KEM-410CCA or some other letter variation at the end. Use this part number to order a replacement laser assembly. There are 4 screws to remove and 1 ribbon cable to disconnect. Use Picture 54 for reference if needed. Lift the laser deck up and out. There are plastic or rubber spacers underneath each screw that need to be saved for re-installation. Once you have a replacement laser deck, reinstall it by seating it back in place remembering to use the spacers underneath each screw. Connect the ribbon cable previously disconnected. Follow the "Drive Reassembly and Testing" section to finish the repair.

Picture 53 (Drive type B laser part number identification)

Picture 54 (Drive type B laser deck removal)

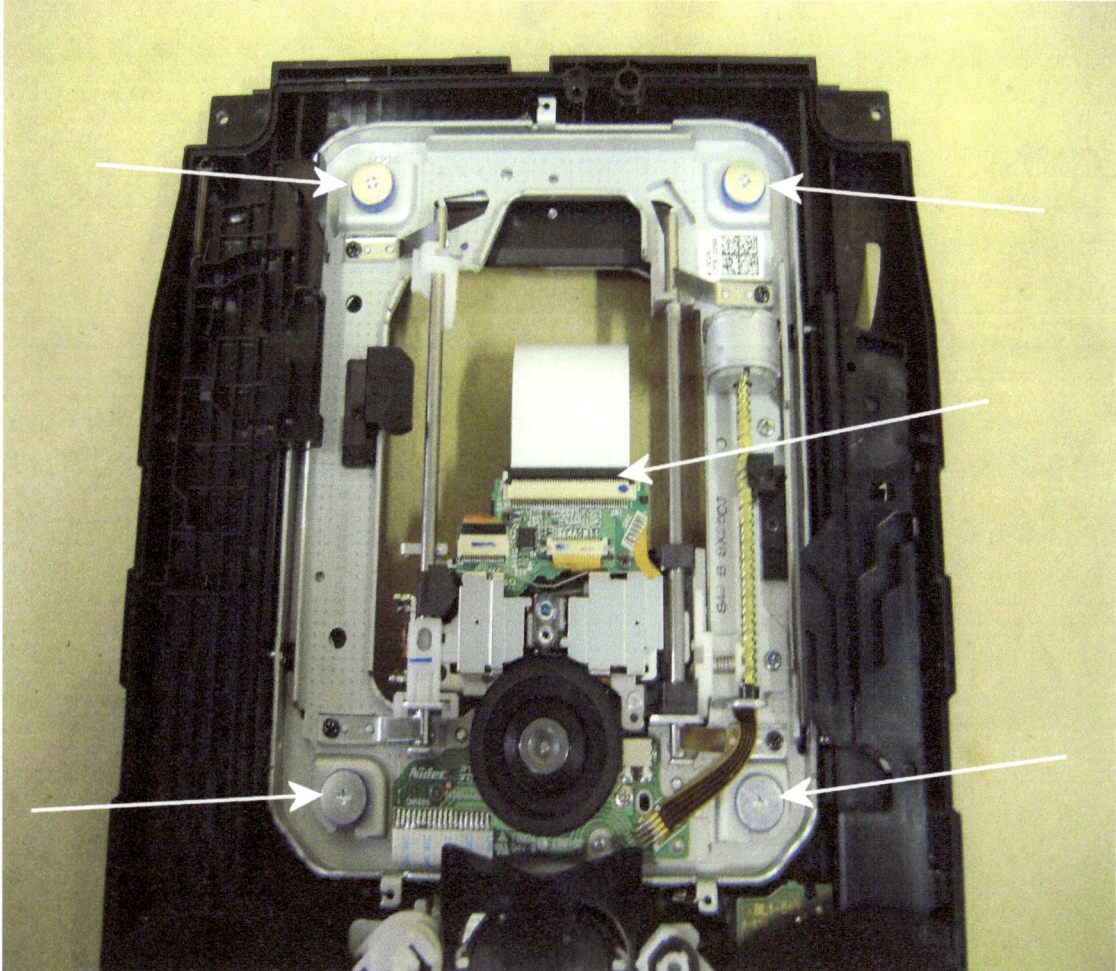

Drive Reassembly and Testing

If previously removed, reinstall the gear system in the previous position pushing on the sides until it snaps back in place. Put the top back on the drive and compare the area pointed to in Picture 55. If it looks like the incorrect one, remove the top again and move the white gear down. Put the top back on and make sure it does not move back up blocking the hole pointed to in the picture. Secure the top of the drive in place with the 2 small black screws and make sure the tab on each side snaps in place. Turn the drive over and reinstall the circuit board if previously removed. Reconnect all 3 cables and use a couple screws to secure the board in place temporarily. Test it for proper operation before complete reassembly. This is to save time as occasionally it takes several repair attempts before the drive works properly. Put the white magnetic disc holder back in the center top of the drive and reconnect the 2 cables from the system to the drive. Power the system on and insert a disc. If the drive does not work properly, first double check all cable connections. If these are correct, revisit the repair sections again until it works correctly. It does take some time and patience! When the drive is working correctly, disconnect it from the system. Place it back inside of the top metal cover. Place the metal bottom over it correctly and secure with the 5 previously used screws. Now place the drive back in the system connecting the 2 cables. Reassemble the system using Chapter 12 for assistance.

Picture 55 (Top gear assembly of drive)

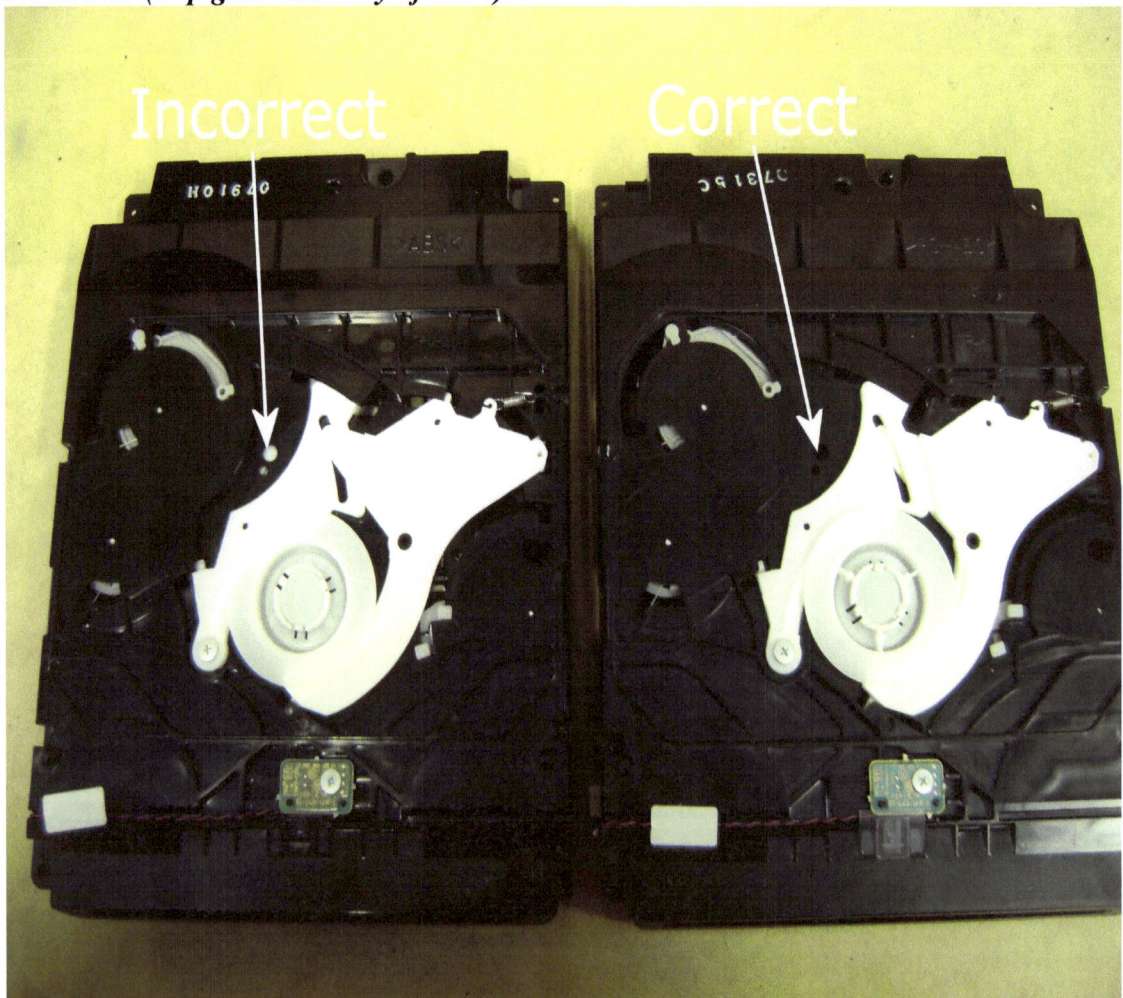

Chapter 5 - Hard Drive Errors and Upgrading

This chapter discusses repair procedures for hard drive issues and upgrading. If the system works fine and a hard drive upgrade is the only thing desired, go to the "Hard Drive Removal and Replacement" section to begin. If there is an error message on the screen that states "hard drive / disc not found or detected", it means that the hard drive is either defective or not properly installed. This error is shown in Picture 56. First, ensure the hard drive is properly installed. The hard drive is located behind the hard drive door. The door is located on the left side of the console if it is sitting horizontally. Picture 57 below identifies the hard drive door. Remove the hard drive door with a small flat-head screwdriver or anything that can fit in the gap on the right side of the door. The hard drive should be seated all the way into the system and the blue mounting screw installed as shown in Picture 58. If the hard drive is installed properly, then it is defective. Please view the "Hard Drive Removal and Replacement" section below to replace the hard drive.

Picture 56 (No Hard Drive found error)

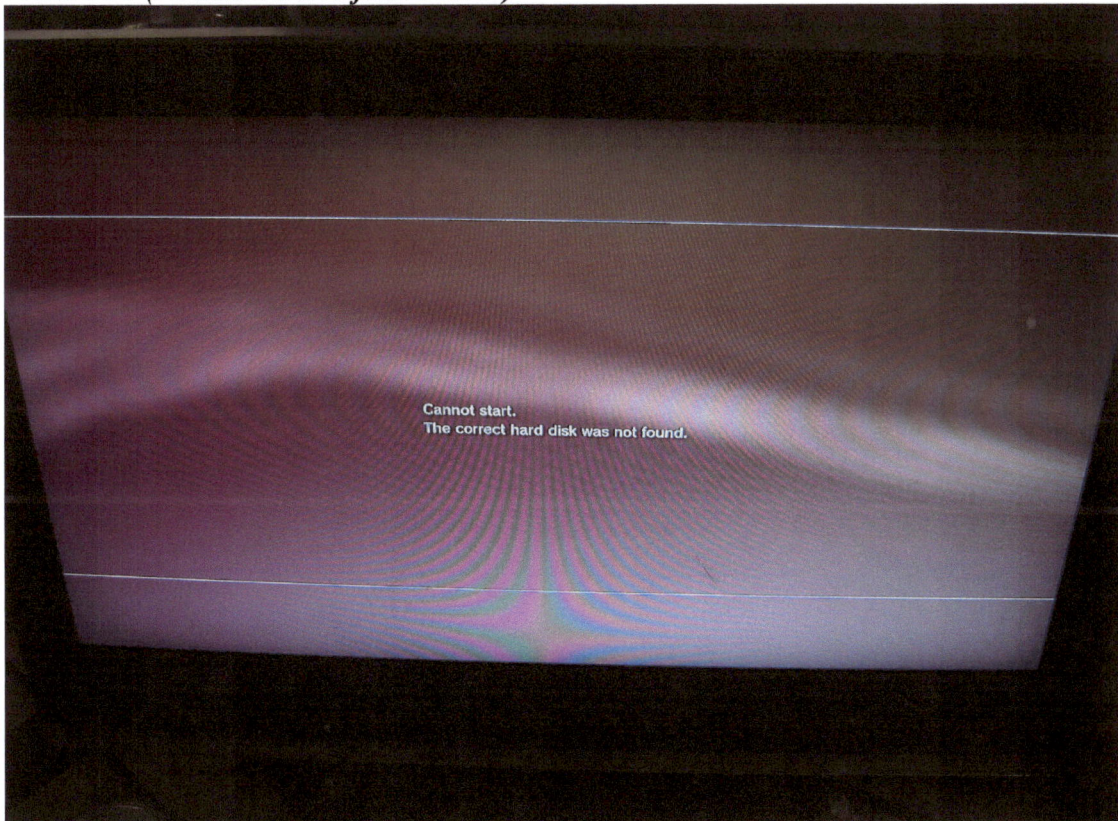

Picture 57 (Hard Drive door location)

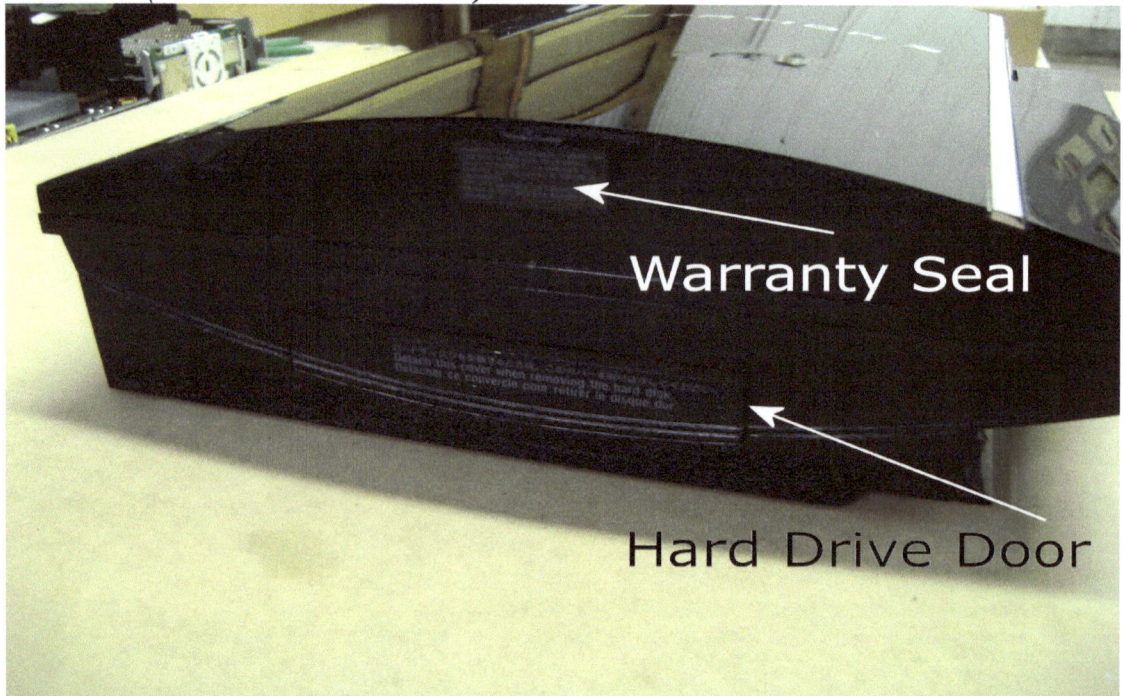

Warranty Seal

Hard Drive Door

Picture 58 (Hard Drive location and door removal)

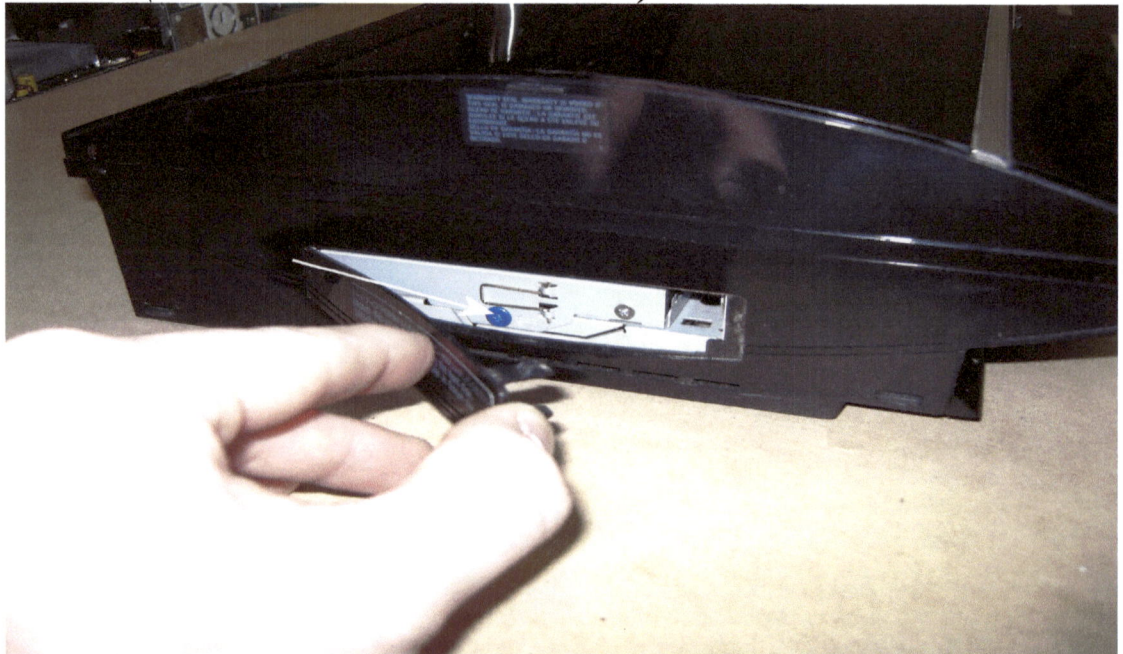

Another common symptom is various error messages like the following: system information and/or software is corrupted, file system is corrupted and needs to be restored, or database needs to be rebuilt. It will also keep looping back to this after completing the process. This issue means the hard drive is defective. Please view the "Hard Drive Removal and Replacement" section for instructions.

If there is a colored background on the screen and the XMB menu never appears, first try the system's "safe mode". The console should be off and the red standby light showing on the front of the console. Push and hold down the power button. The system will turn on and shortly turn back off. Once the system turns off (green light changes to red again), let go of the power button. Immediately, push and hold the power button down again. The system will turn on, beep once, and then beep twice. As soon as it beeps twice, release the power button. A black screen will appear prompting you to connect a controller. After that, a menu screen should appear like in Picture 59.

Picture 59 (Safe mode screen)

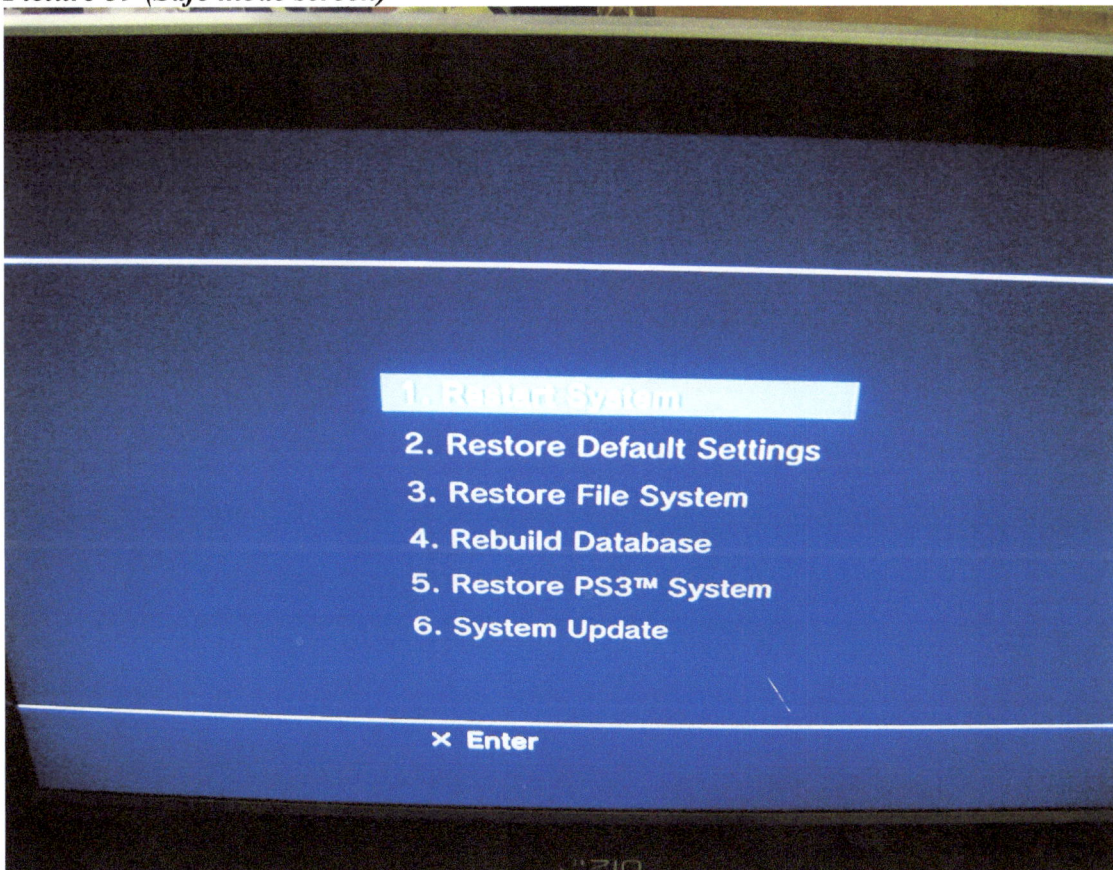

If the system will not complete a software update, first complete the "System Software Installation" section. If this does not fix the problem, confirm that the disc drive works properly. We have found a defective or missing (the 2 large silver ICs) drive will cause it not update. If it will play all movies and games in it, the disc drive is not the problem. If the drive is not working correctly, visit Chapter 5 to repair it before updating the system. Once the drive is working, follow the "System Software Installation" section in this chapter to update software. If the drive is working correctly, the hard drive is defective. Continue to the next section for repair procedures.

- Hard Drive Removal and Replacement

Remove the hard drive door to expose the hard drive. Picture 57 on the previous page can be used as a reference. After the hard drive is exposed, remove the big blue screw that secures the hard drive in place. There are 2 different styles of hard drive caddies or mounts. There is a slight difference in removal depending on which style it is. Identify which model you have by looking on the white sticker at the back of the console. If your system is model CECHA01, CECHB01, CECHE01 or CECHG01, grab the metal handle; pull it to the right and then out. If your system is model CECHH01, CECHP01, CECHK01 or CECHL01, grab the metal tab you removed the screw from and pull it out. Now with the hard drive caddy out, there are 4 small screws on the hard drive mount / caddy to remove that will release the hard drive. These screws are shown in Picture 60 and 61. Now slide the hard drive out of the mount.

Picture 60 (Screws on hard drive caddy for removal part 1)

Picture 61 (Screws on hard drive caddy for removal part 2)

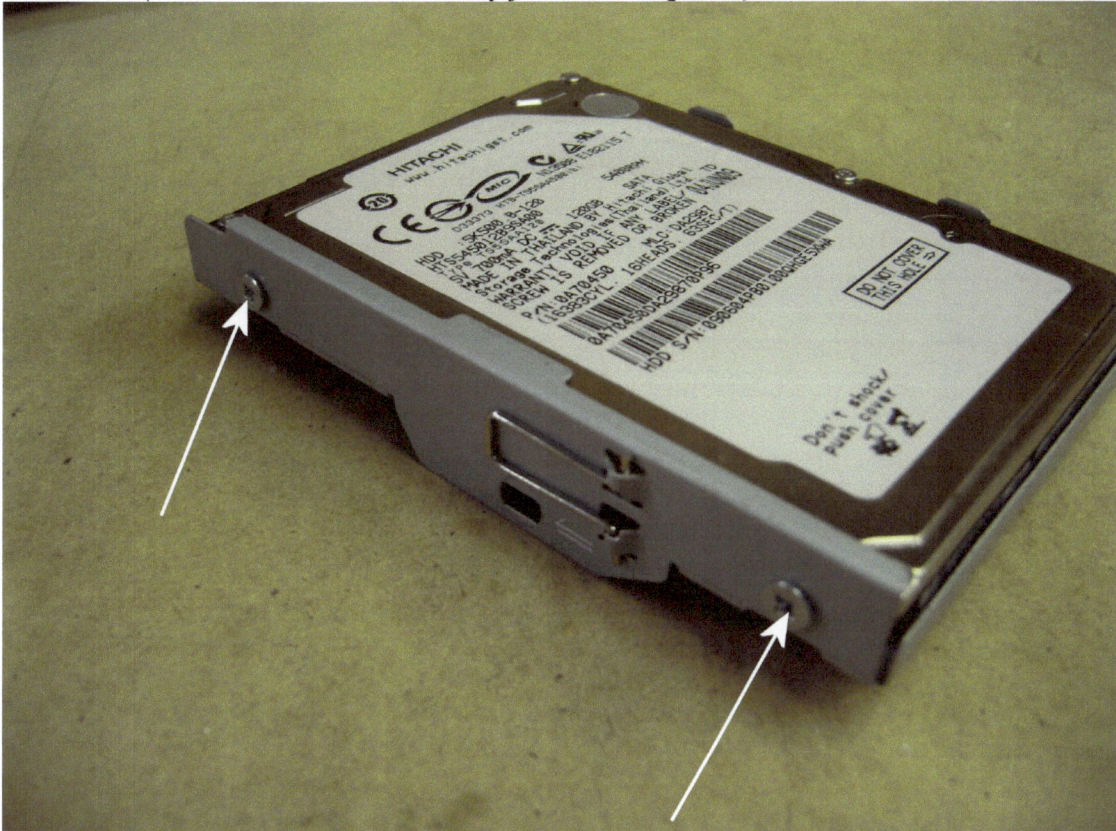

At this point, decide whether or not to upgrade the hard drive size (capacity). There are many places on the internet that sell hard drives that are compatible. To reinstall, slide the new hard drive into the caddy / mount and secure the 4 screws previously used. Push the hard drive caddy / mount back into the system firmly. Reinstall the blue screw to hold the hard drive caddy / mount in place. Put the hard drive door back on. Now continue to the "System Software Installation" to complete the repair.

- System Software Installation

The newest system software for the system is found on Sony's website under the support section. Please carefully read and follow the instructions on how to download and prepare the file for installation. It is highly recommended putting the system software on a USB flash drive for installation. The USB drive should contain a folder named "PS3" with the system software update file inside the folder. Once this is completed, connect the USB flash drive and turn the system on. If the hard drive has not been replaced and is just being updated, choose the "System Update" option in the menu screen. Choose to update from the attached media and not the internet.

If a new hard drive is being installed, a black screen will appear saying that system software was not found and needs to be installed. The select and start buttons on the controller need to be held until the screen changes. After a few minutes, it will confirm the USB flash drive with the system software. Hold select and start buttons again until the installation begins. If the system cannot find the software, confirm the USB drive connected properly and the system software is correctly placed on the USB flash drive. The installation process will take several minutes. When completed, the system will restart and the XMB (menu) should appear on the TV.

Chapter 6 - Video and Audio Issues

If the symptom is video and/or audio problems, first eliminate any sync issues between the system and the TV. To do this, you will need to access a certain system menu often referred to as "safe mode". The console should be off and the red standby light showing on the front of the console. Push and hold down the power button. The system will turn on and shortly turn back off. Once the system turns off (green light changes to red again), let go of the power button. Immediately, push and hold the power button down again. The system will turn on, beep once, and then beep twice. As soon as it beeps twice, release the power button. A black screen will appear prompting you to turn on the controller. After that, a menu screen should appear like in Picture 62. Choose option 3 (Restore File System) and agree to the on-screen prompts. After it is complete, the system will restart to the XMB menu like normal. If the system menu screen did not appear, the problem is more complex. This problem is usually caused by poor solder connections underneath the CPU and or the GPU. We will have to use a reflow process on the motherboard to fix this. Please visit Chapter 10 for the repair procedure.

Picture 62 (System menu screen)

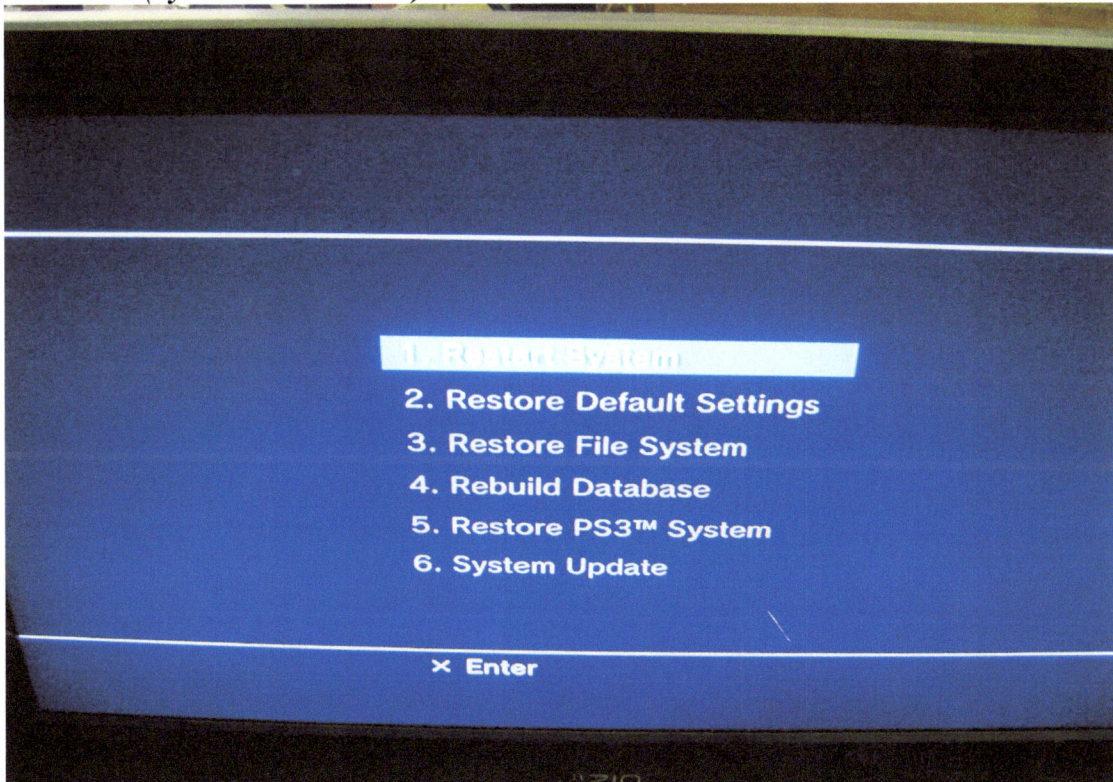

Chapter 7 - Power and Media Card Reader Issues

If the system will not read or write to a media card, CECHA01 and CECHE01 models only, the issue is a defective media card reader. Use Chapter 3 to disassemble the system and follow the section on removing the media card board. Once removed, there is a part number on the board inside of the metal covers. The part number should be CMC-001. Use this part number to find a replacement board. Many websites sell this laser assembly. Beware of purchasing a generic or aftermarket part. Reinstall the replacement board by reconnecting the ribbon cable on the bottom and securing in place with the 2 previously used screws. Now reassemble the system in reverse order of disassembly and the problem is fixed.

This section will cover when a system is "dead" or unresponsive system when the power button is pushed. Another symptom is that there will not be a red (standby) light on the front of the console when power is applied. Picture 63 shows the red standby light on the front of the console. The power supply is the most common part that causes this issue. Refer to Chapter 3 to disassemble the system and remove the power supply. The power supply has a model / part number on the bottom of it starting with APS or EADP. That number is what will be used to purchase a replacement. Many websites sell this laser assembly. Beware of purchasing a generic or aftermarket part.

Picture 63 (Red standby light on front of console)

56

To install the replacement power supply, seat it back on the motherboard making sure to fit it onto the 2 gold prongs as shown in Picture 64. Reconnect the 2 cables previously disconnected along with reinstalling screws. At this point, reconnect the AC power cable and turn the power switch on. Check for the red standby light on the front of the console. If the red light appears, reconnect the video cable from the TV to the system and turn it on to confirm everything is working properly. Reassemble the system in reverse order of disassembly and use Chapter 12 for help if needed. If it is not turning on after changing the power supply, this means the issue is with the solder connections on the CPU and GPU. To fix this problem, we will have to use a reflow process on the motherboard. Please visit Chapter 10 for instructions on this procedure.

Picture 64 (Power supply re-installation)

Chapter 8 – Overheating Issues

If a message appears on the screen stating the system is overheating and will shut down, this may be a simple problem to fix. First, try using a can of compressed air to clean out all the air vents that are located around the console. You also may reconsider where you have the console placed. It is important to proper ventilation room for the system to cool properly. It should not be in an enclosed cabinet. If the system is still overheating after this, it will have to be opened up. Visit Chapter 3 for detailed instructions on how to disassemble the system completely down to the motherboard. During this process, it is very important to clean the complete system out with compressed air as it is taken apart.

Once the system is completely disassembled down to the motherboard, clean the old thermal compound off of the CPU and GPU (the 2 large silver ICs) as in Picture 65. Also, completely clean the compound off the heat sink / fan assembly as in Picture 66. Reapply a high quality thermal compound on the CPU and GPU as shown in Picture 67. Spread the compound over the complete IC. Place the heat sink / fan assembly back on top of the motherboard. Now overheating issues should be solved. Use Chapter 12 to help reassemble the system.

Picture 65 (Cleaning thermal compound off of the CPU and GPU)

Picture 66 (Cleaning of heat sink / fan assembly)

Picture 67 (Reapplying thermal compound on ICs)

Chapter 9 - Shutdown Issues

If the console will turn on and soon after shuts back down, the issue is commonly described as the "Yellow Light of Death" or YLOD. It is characterized by a yellow light flashing on the front of the console before shutting down as shown in Picture 68. This issue occurs when there are problems with the solder connections underneath the CPU and GPU, which are located on the motherboard. This issue is repaired by using a reflow process on the motherboard to fix any of these solder connections. Please visit Chapter 10 for details on the repair procedure.

Picture 68 (Yellow light on front of console)

Chapter 10 - Motherboard Reflow Process

The reflow process attempts to repair any poor solder connections that are underneath the ICs (Integrated Circuits) on the motherboard. These poor solder joints are often referred to as "cold solder joints". A cold solder joint is a weak or unreliable solder connection from the IC to the motherboard. This often occurs when the motherboard begins to flex or warp after heating up and cooling down from years of use. To repair this, the system needs to be disassembled down to the motherboard. View Chapter 3 for detailed disassembly instructions down to the motherboard. After this, we will need to clean off the old thermal compound on the CPU and the GPU (the 2 large silver ICs) using a non-corrosive cleaner with cotton swabs or cloth like in Picture 69. Also, clean off all thermal compound from the heat sink / fan assembly as shown in Picture 70. Remove the backup battery from the motherboard like in Picture 71. Remove all the insulators that are on top of or around ICs as shown in Picture 72. You will need a heat gun for this process. A heat gun is inexpensive and readily available at just about any hardware or home improvement store. A hair dryer cannot be used as it does not achieve the high temperatures needed for the reflow process (750°F). Using a circuit board reflow / rework station is the preferred method versus a heat gun, but that type of equipment is obviously not available to most people.

Picture 69 (Cleaning thermal compound off of CPU and GPU)

Picture 70 (Cleaning of heat sink / fan assembly)

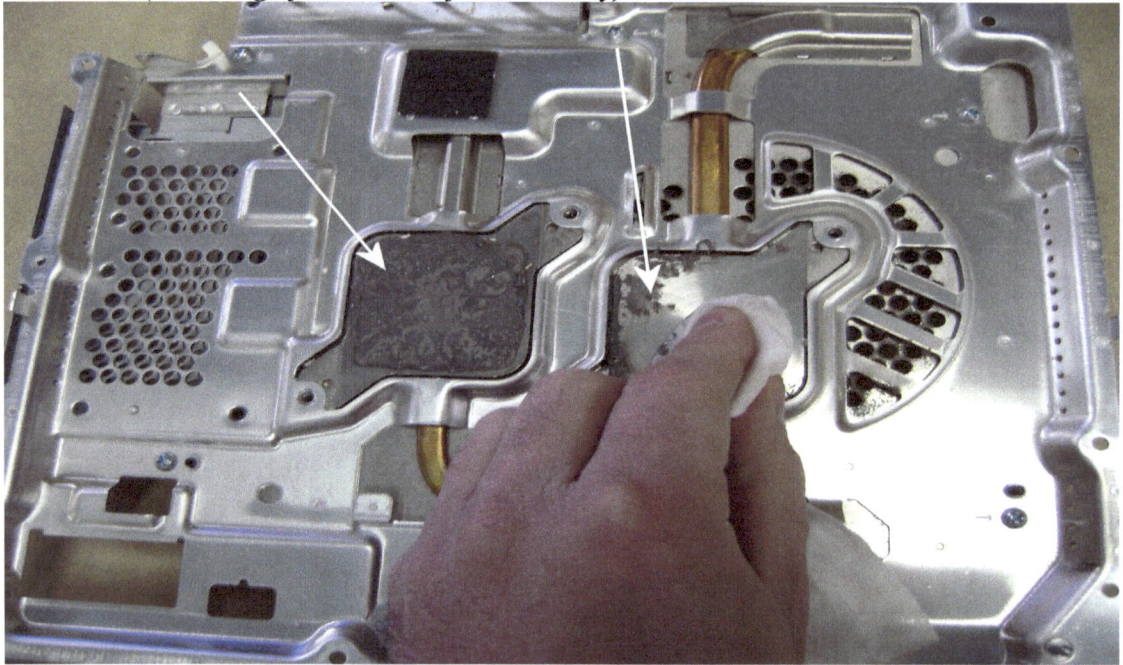

Picture 71 (Removal of backup battery)

Picture 72 (Removal of insulators on and around ICs)

The first step of the reflow process is to pre-heat the motherboard. Place the board so it is sitting flat. You may want to wear something to protect your hands from the heat. Set the heat gun to around 750°F. Keep it about 4 inches away from the board and move it constantly around for 1 minute on each side (bottom and top). Remember to keep the heat gun constantly moving. Never keep the heat gun still as it can cause damage to components. After this, focus the heat gun directly on the CPU and GPU holding it about 2 inches away for 1 minute on each IC. Please remember to keep the heat gun constantly moving around the IC. This process is shown in Picture 73. Once this has been completed, turn the heat gun off and leave the motherboard untouched for 15 minutes so it can cool down. This is a very important part of the process. Moving the board around at this point can cause damage that may be irreparable. Once the motherboard has completely cooled, reapply a high quality thermal compound on the CPU and GPU like in Picture 74. There should be enough thermal compound to thinly cover the complete IC. Reassemble the system using Chapter 12 as reference if needed. Reconnect the power cable and turn the power switch to on. There should be a red standby light on the front of the console now. Reconnect the video cable from the TV and turn the console on. If the problem is not fixed, repeat the reflow process increasing to 2 minutes on pre-heating the board and the amount of time focused on the CPU and GPU respectively. If it is still not working properly after completing this, unfortunately the system is apart of the very small percentage that will not be repairable.

Picture 73 (Reflow process)

Picture 74 (Reapplying thermal compound to CPU and GPU)

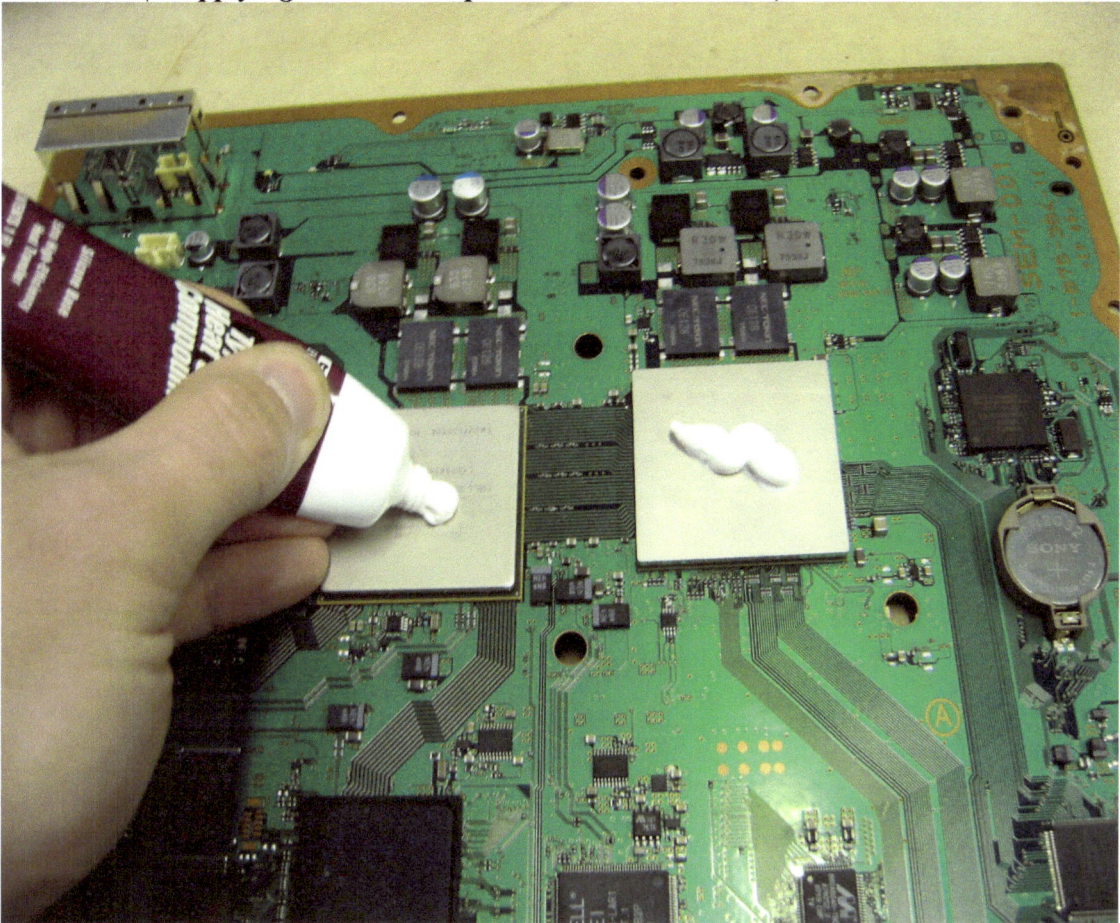

Chapter 11 - Internet / Network and Controller Connection Issues

Most internet or online network connectivity issues are not hardware related with the system hardware. If you are not able to connect to PlayStation® network, it is usually due to a temporary problem with their servers or your wireless internet connection to the console. If you are experiencing any internet connection issues, go to the "Network Settings" in the XMB menu and double check the internet connection settings like in Picture 75. After this, use the "Test Network Connection" option to ensure the system is connected to the internet like in Picture 76. If it fails this test, look at the error that is received. Typically, the internet connection is the problem. Check your internet connection devices such as cable modem or router.

Picture 75 (Internet connection settings in the XMB menu)

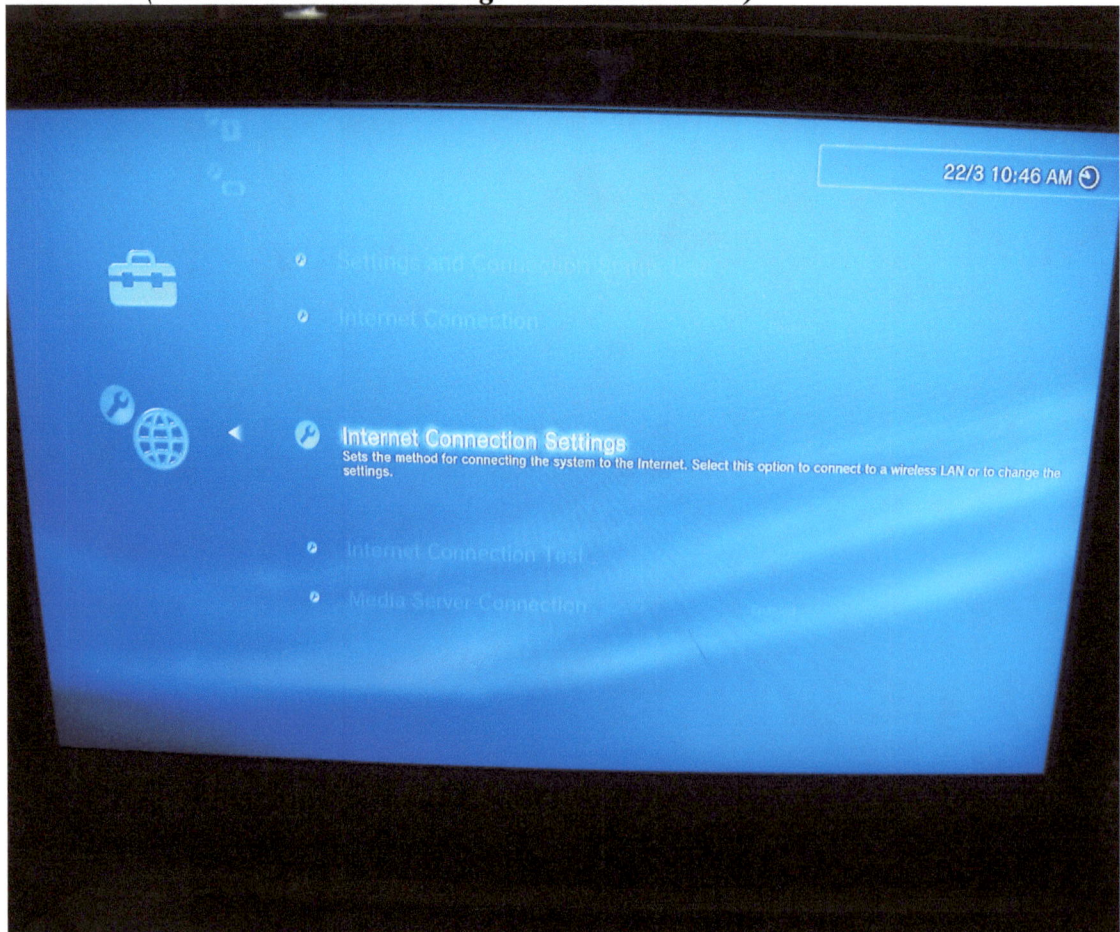

Picture 76 (Internet connection test)

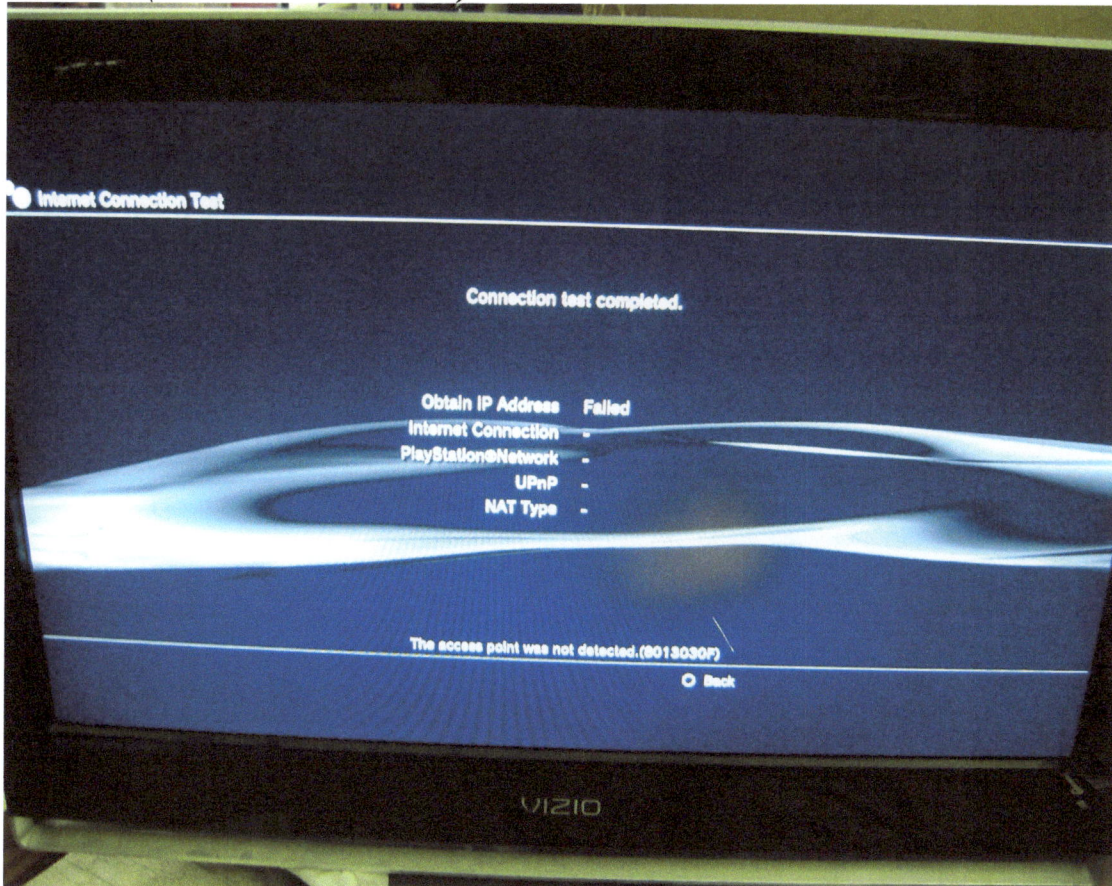

Check the error received when signing in. Ensure your login information is correct and that your user ID has not been banned. If you have any user ID issues including being banned, contact Sony. The system used is banned along with the user ID if this happens. Once all other scenarios are eliminated as the problem with your internet connectivity, the Wi-Fi / Bluetooth board inside of the console is defective and needs to be replaced.

Wi-Fi / Bluetooth Board Replacement

Visit Chapter 3 to disassemble the system and remove the Wi-Fi / Bluetooth board. Only the main board that is in front of the power supply should be replaced. Once this board has been removed, you will need a part number off the board. There should be a part number on the board beginning with CWI. Use this number to find a replacement.

Many websites sell this laser assembly. Beware of purchasing a generic or aftermarket part. Reinstall the replacement board in reverse order of disassembly. Ensure both cables are connected and secure the board in place with the previously used screws. View Chapter 12 for details on reassembling the system.

　　　If the issue with the system is problems with a controller connecting or working properly, there are a couple issues that can cause this. Commonly, the controller is causing the issue and not the system. Ensure that the controller battery is charged first by connecting the USB cable (from controller to the system) for charging. After ensuring a charge, reset the controller. Push and hold the reset button on the back of controller with a paper clip for 10 seconds. Use Picture 77 for reference. After this, reconnect the USB cable and turn the controller on. If the controller will still not connect, try a different controller to find out if the system is the issue.   If another controller will connect to the system, the previous controller was the issue. If you cannot connect any controller to the system, the Wi-Fi / Bluetooth board is defective. Look at "Wi-Fi / Bluetooth Board Replacement" on the previous page to assist in this repair.

Picture 77 (Controller reset button)

Chapter 12 – System Reassembly

This chapter will start from complete system disassembly back to a completely reassembled system. Seat the motherboard back on top of the fan / heat sink assembly as shown in Picture 78. Make sure to align the AV connections on the board to the holes on the fan / heat sink assembly. Reinstall the top metal cover onto the motherboard aligning the tabs as shown in Picture 79. Place the metal clamps over the ICs and reinstall the screws shown in Picture 80. Turn the motherboard over and reconnect the fan cable as shown in Picture 81. Reinstall cover over the AV jacks like in Picture 82. Sit the motherboard back into the bottom plastic case of the console like in Picture 83. Reinstall screws in all the holes pointed to in Picture 84. There are 2 unique screws that are placed next to their corresponding holes in the picture also. Remember to secure the ground wire (yellow and green wire) with the corresponding screw pointed to the picture.

Picture 78 (Seating motherboard back onto the fan / heat sink assembly)

Picture 79 (Top metal cover placement on motherboard)

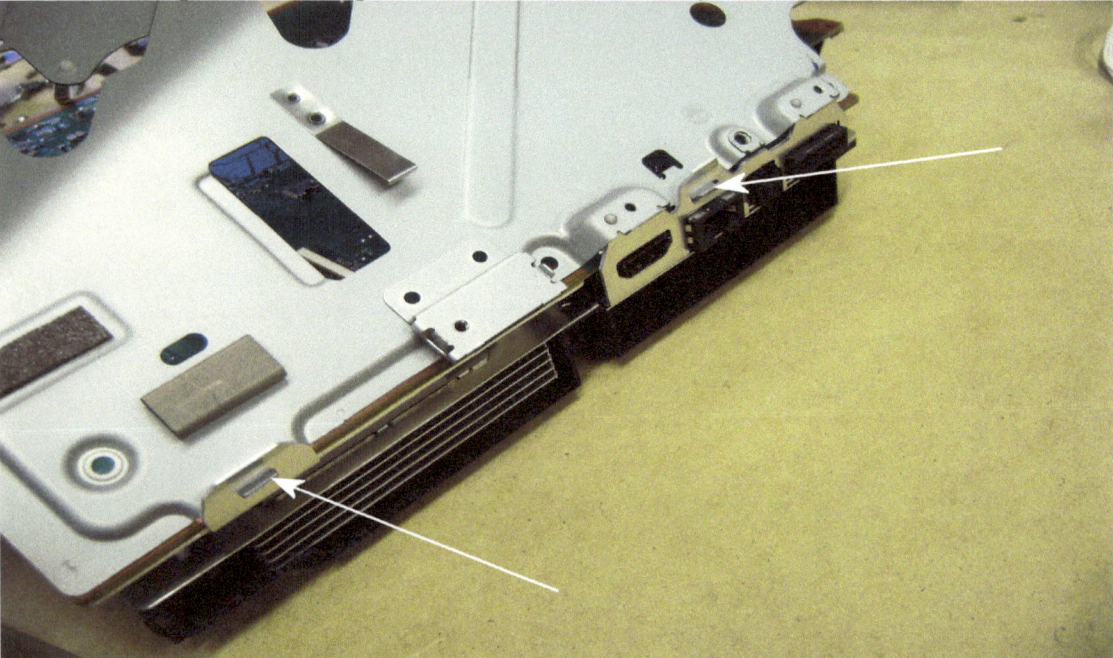

Picture 80 (Metal clamps and screw installation)

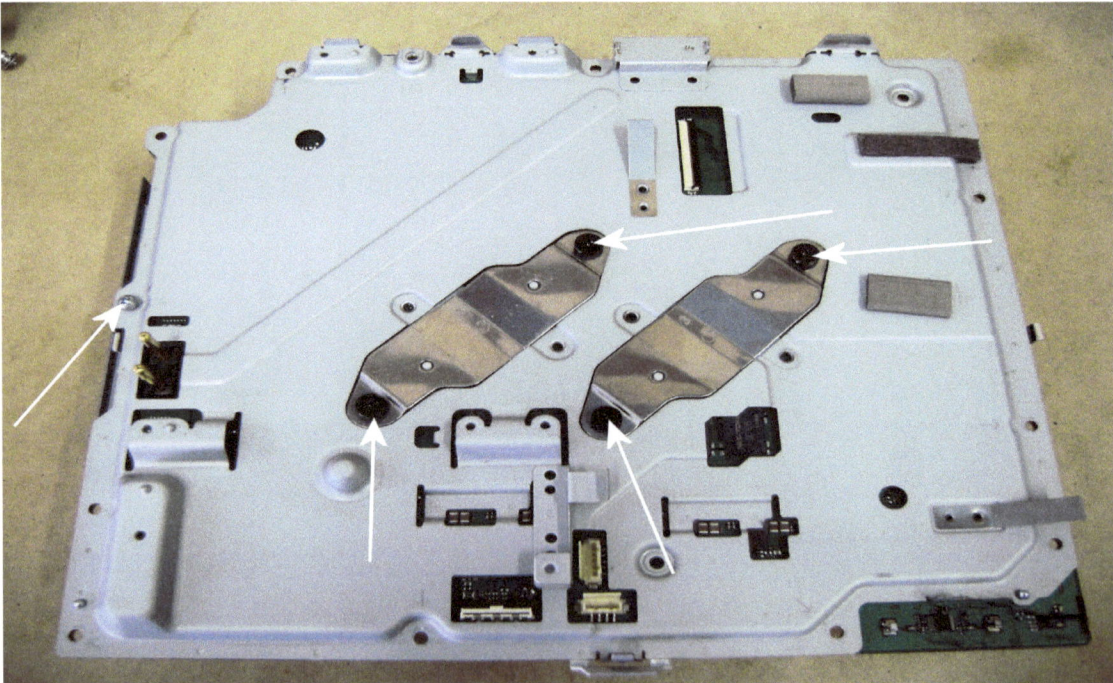

Picture 81 (Reconnecting fan cable to motherboard)

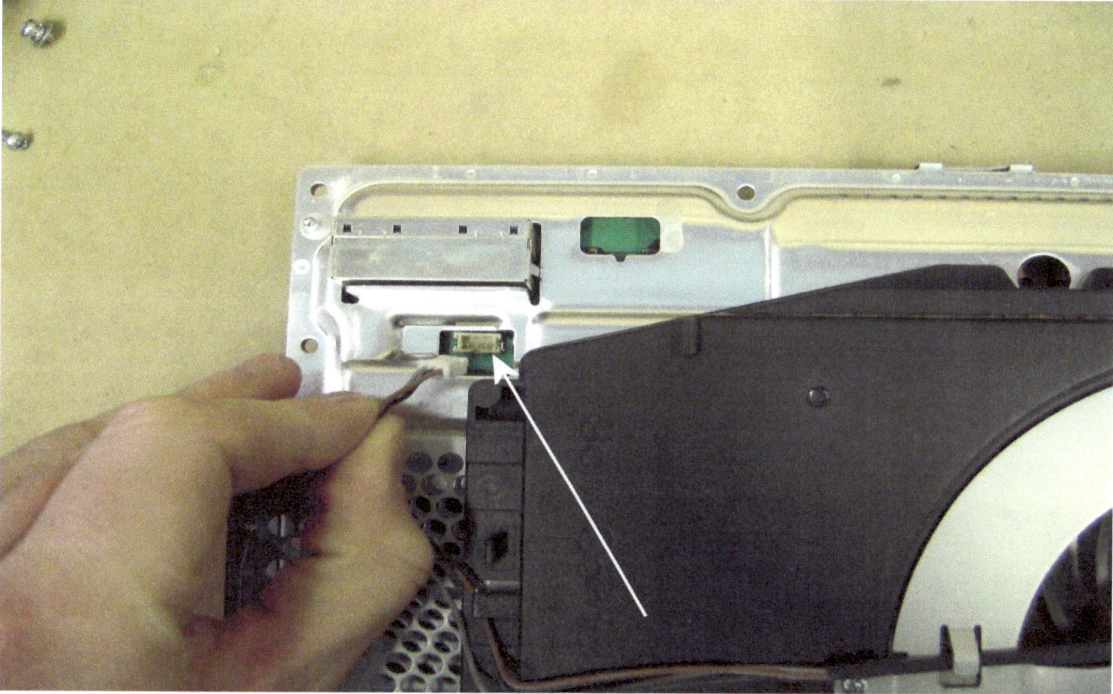

Picture 82 (AV cover re-installation)

Picture 83 (Placing motherboard into bottom case of console)

Picture 84 (Screw installation to secure motherboard)

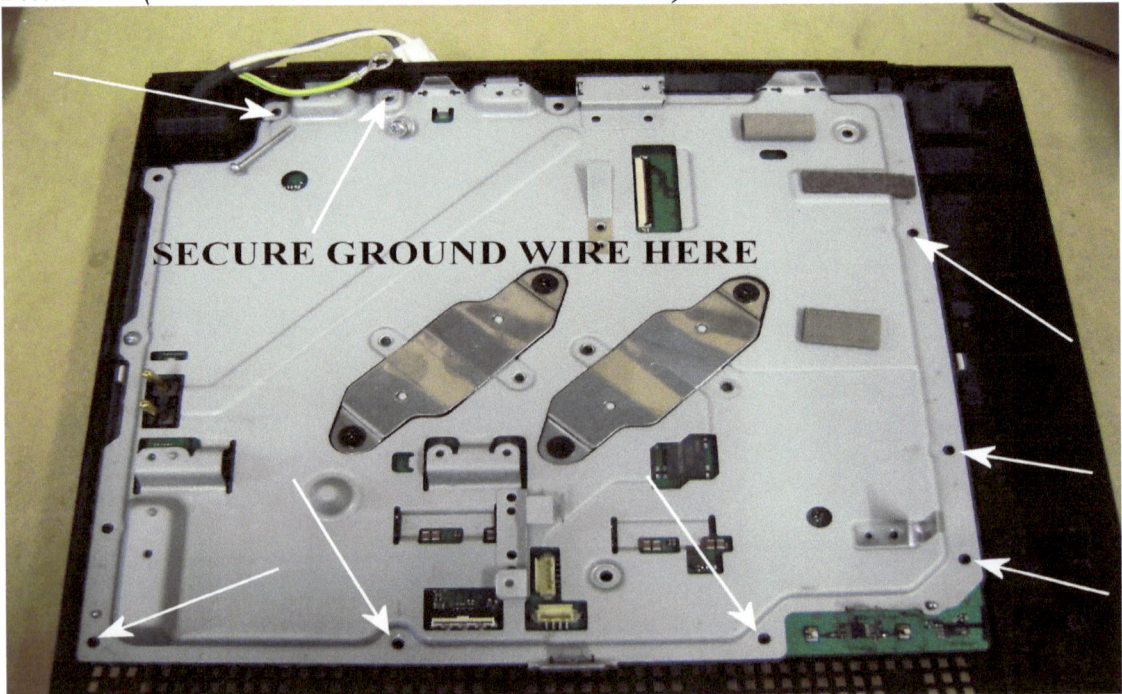

SECURE GROUND WIRE HERE

Reinstall the 4 cables that were previously removed from the motherboard. Reinstall the hard drive by pushing it back into place. Secure it with the large blue screw previously used. Put the hard drive door back on by inserting it from the left and it will snap in place. Seat the power supply back on the motherboard. Reconnect the 2 cables and 5 screws to secure it in place as shown in Picture 85. Reconnect the 2 cables to the disc drive (1 ribbon cable underneath and 1 cable on the left side) as shown in Picture 86 and 87. Reinstall the secondary Wi-Fi board as shown in Picture 88. Reinstall the main Wi-Fi board by securing it with 2 screws. Reconnect the 2 cables to it like in Picture 89. Place the top cover back on the console by reconnecting it from the front first as in Picture 90. Reinstall the 7 screws to secure the top case of the console. Remember the screw in the top right corner is shorter than the others. Use Picture 91 for reference. Reinstall the reflective top cover by seating it on and sliding it forward like in Picture 92. Remember if your system had a metal tab like in Picture 93 when originally removing it; reinstall that tab before placing the cover on. Reinstall the T10 Torx screw by placing it through the hole on the left side of the console like in Picture 94. After securing the top cover on, push the black plastic insert back in place. Reconnect the power and AV cable and reassembly is complete!

Picture 85 (Reinstalling the power supply)

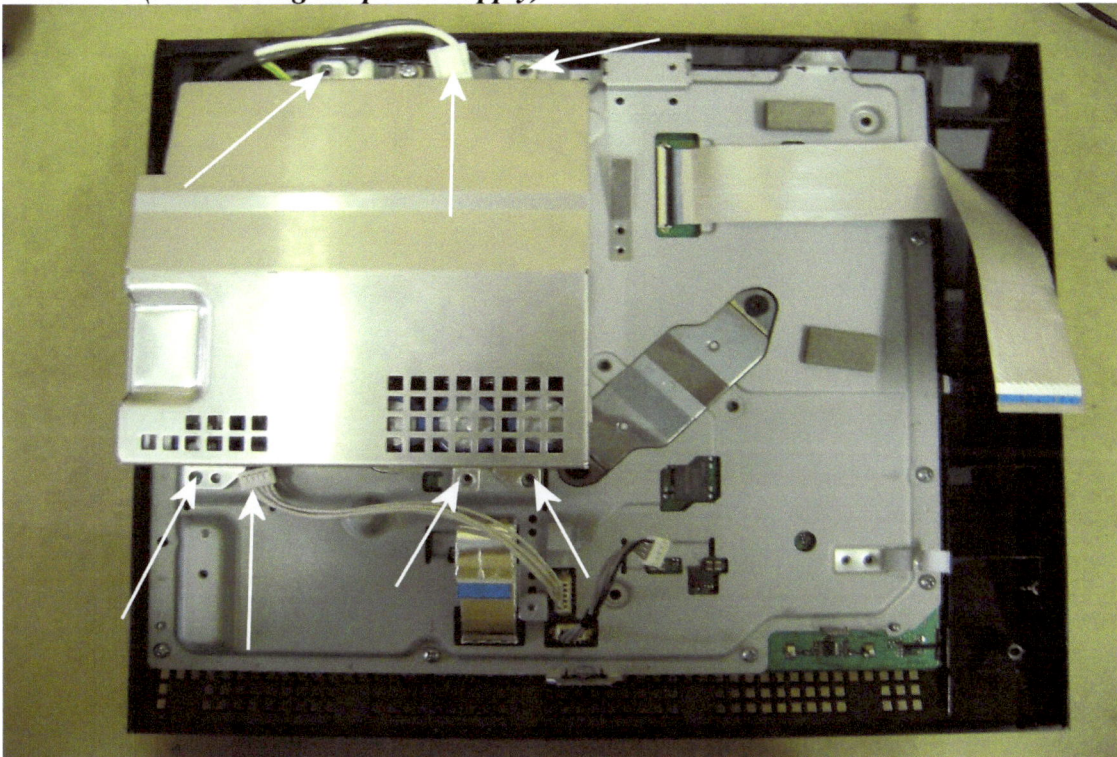

Picture 86 (Reinstalling disc drive)

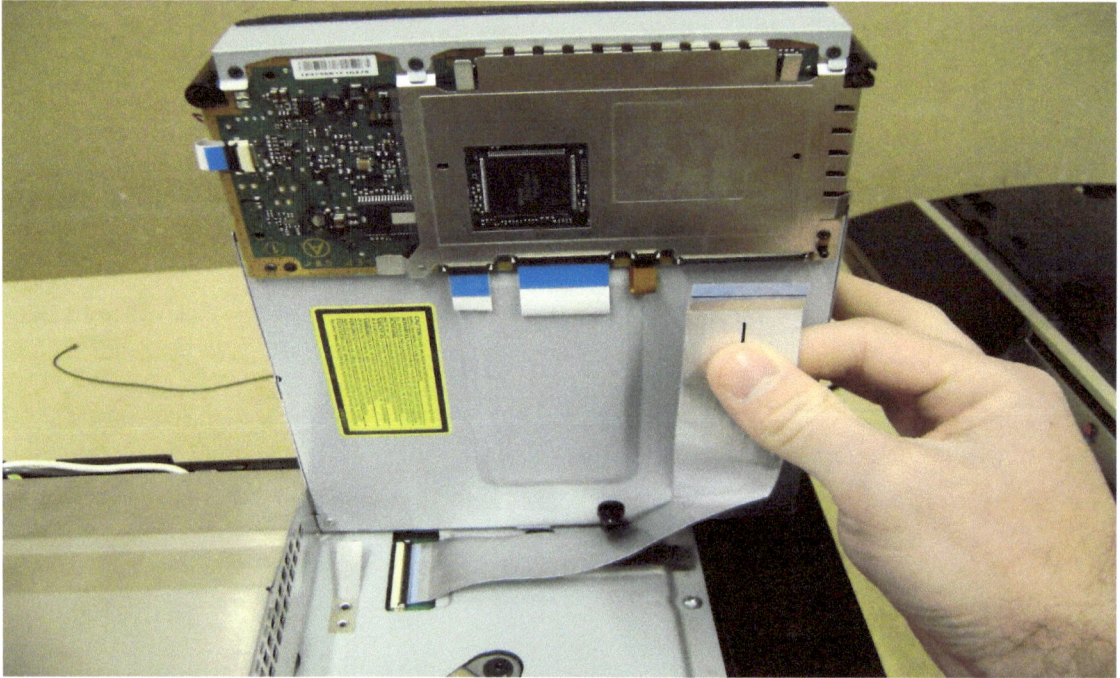

Picture 87 (Reinstalling disc drive continued)

Picture 88 (Wi-Fi secondary board re-installation)

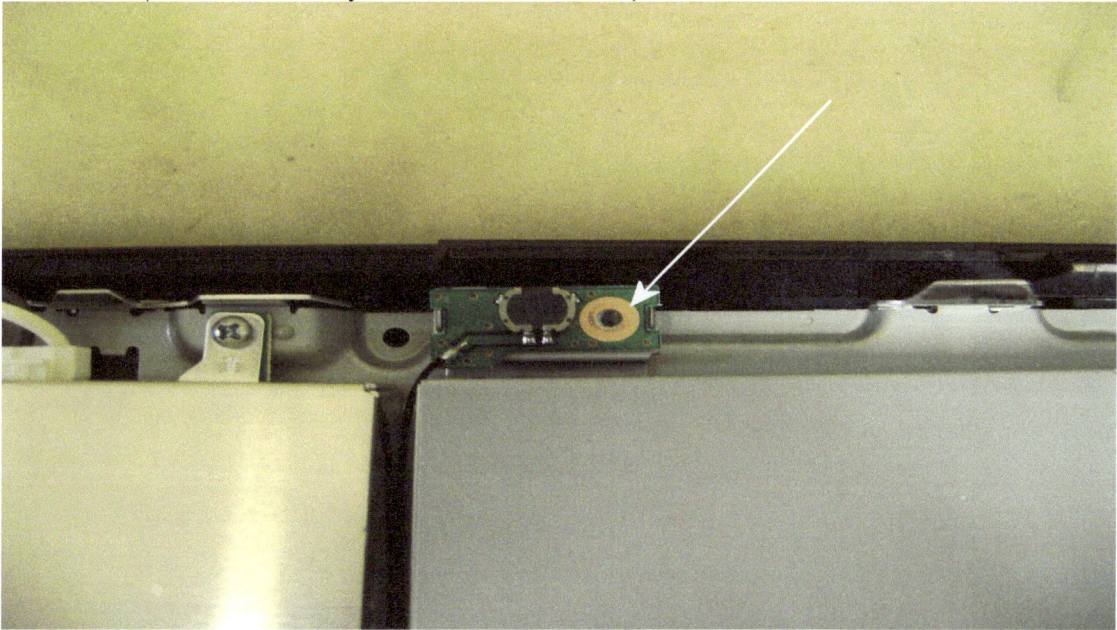

Picture 89 (Wi-Fi main board re-installation)

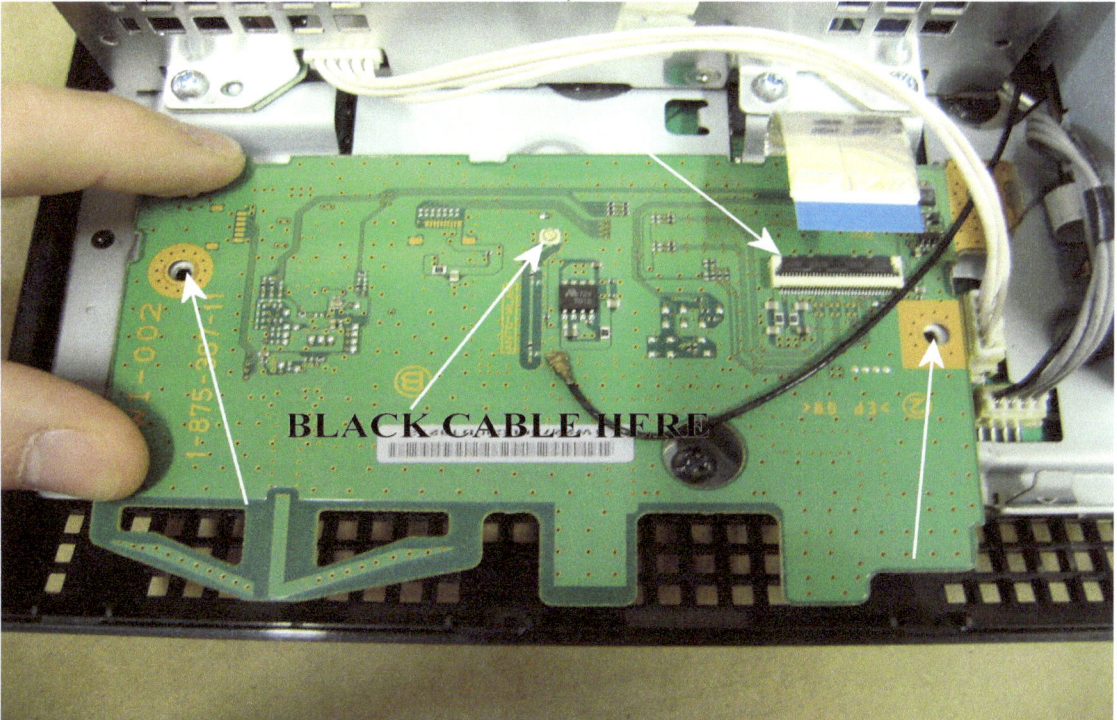

BLACK CABLE HERE

Picture 90 (Top case installation)

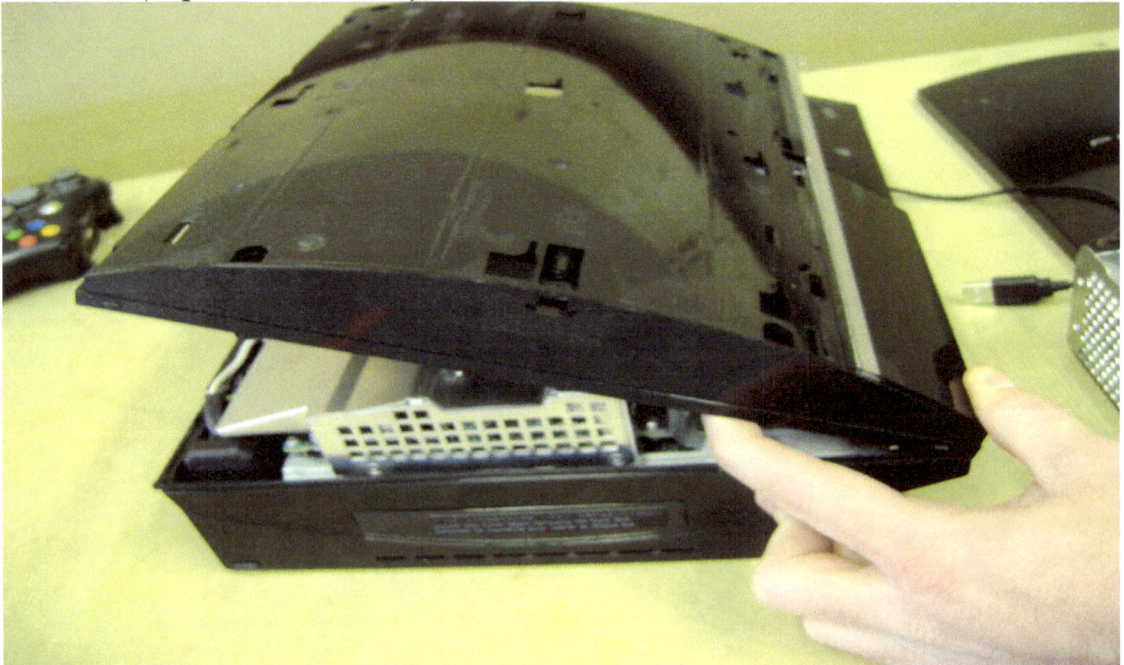

Picture 91 (Reinstalling screws to secure top case)

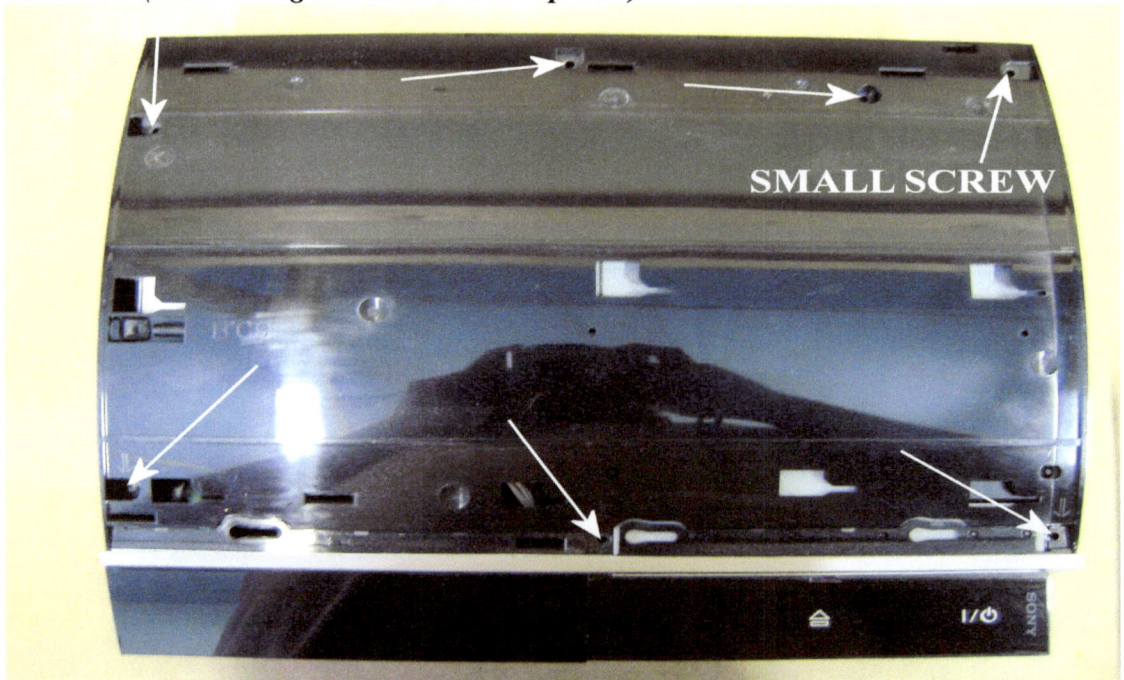

SMALL SCREW

Picture 92 (Reinstalling top cover)

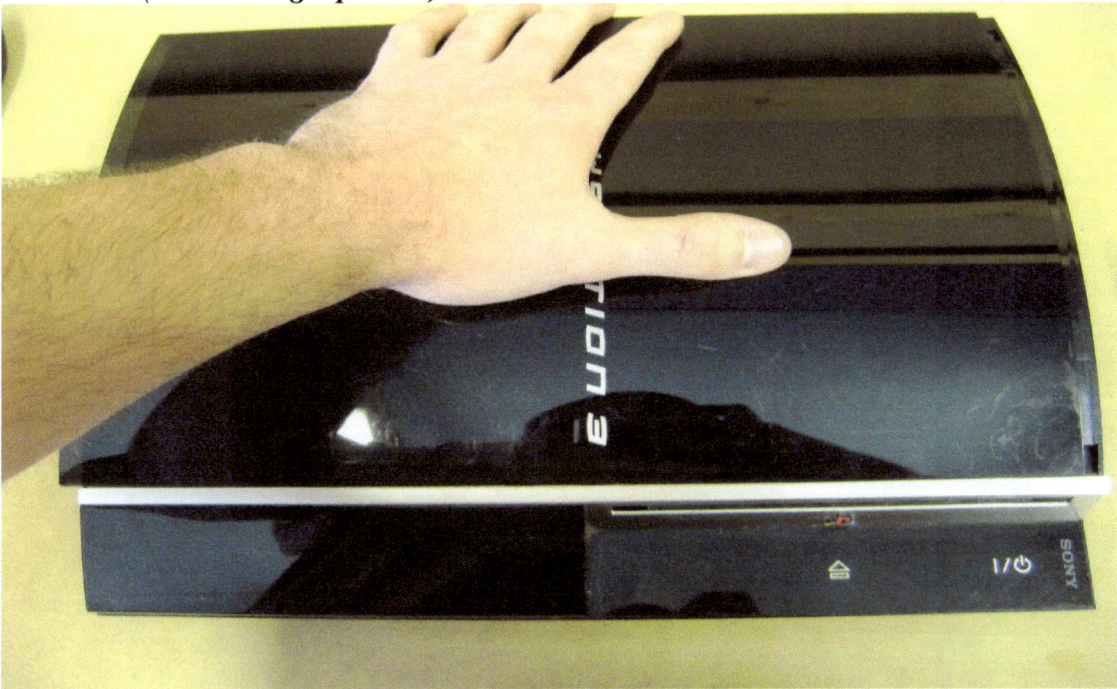

Picture 93 (Metal tab re-installation if originally used)

Picture 94 (Securing the top cover in place)

www.ingramcontent.com/pod-product-compliance
Lightning Source LLC
Chambersburg PA
CBHW052052190326
41519CB00002BA/196